EVERY STUDENT CAN LEARN MATHEMATICS

MATHEMATICS Instruction & Tasks in a PLC at Work®

SECOND EDITION

Mona Toncheff
Timothy D. Kanold
Sarah Schuhl
Bill Barnes
Jennifer Deinhart
Jessica Kanold-McIntyre
Matthew R. Larson

A Joint Publication With

555 North Morton Street
Bloomington, IN 47404
800.733.6786 (toll free) / 812.336.7700
FAX: 812.336.7790

email: info@SolutionTree.com
SolutionTree.com

Visit **go.SolutionTree.com/MathematicsatWork** to download the free reproducibles in this book.

Printed in the United States of America

Library of Congress Cataloging-in-Publication Data

Names: Toncheff, Mona, author. | Kanold, Timothy D., author. | Schuhl, Sarah, author. | Barnes, Bill, author. | Deinhart, Jennifer, author. | Kanold-McIntyre, Jessica, author. | Larson, Matthew R., author.
Title: Mathematics instruction and tasks in a PLC at Work / Mona Toncheff, Timothy D. Kanold, Sarah Schuhl, Bill Barnes, Jennifer Deinhart, Jessica Kanold-McIntyre, Matthew R. Larson.
Description: Second edition. | Bloomington, IN : Solution Tree Press, [2024] | Includes bibliographical references and index.
Identifiers: LCCN 2023012432 (print) | LCCN 2023012433 (ebook) | ISBN 9781958590652 (paperback) | ISBN 9781958590669 (ebook)
Subjects: LCSH: Mathematics--Study and teaching. | Mathematics teachers--Training of.
Classification: LCC QA11.2 .G7427 2024 (print) | LCC QA11.2 (ebook) | DDC 510.71/2--dc23/eng20230711
LC record available at https://lccn.loc.gov/2023012432
LC ebook record available at https://lccn.loc.gov/2023012433

Solution Tree
Jeffrey C. Jones, CEO
Edmund M. Ackerman, President

Solution Tree Press
President and Publisher: Douglas M. Rife
Associate Publishers: Todd Brakke and Kendra Slayton
Editorial Director: Laurel Hecker
Art Director: Rian Anderson
Copy Chief: Jessi Finn
Senior Production Editor: Suzanne Kraszewski
Proofreader: Charlotte Jones
Text and Cover Designer: Abigail Bowen
Acquisitions Editor: Hilary Goff
Assistant Acquisitions Editor: Elijah Oates
Content Development Specialist: Amy Rubenstein
Associate Editor: Sarah Ludwig
Editorial Assistant: Anne Marie Watkins

Acknowledgments

The new and enhanced chapters for this second edition of *Mathematics Instruction and Tasks in a PLC at Work* are based on the deep learning and professional development our author team and Mathematics at Work™ associates have led since that book's publication in 2018. We have been privileged to work across North America with mathematics educators using protocols and resources from the first edition. Questions, insights, and feedback from readers and users of our Mathematics at Work materials are revealed in the examples and revisions designed to strengthen a teacher team's intentional lesson design and implementation of daily mathematical tasks.

A special thank-you to Jeff Jones, Douglas Rife, Todd Brakke, and Rian Anderson. Special thanks to the incredible Suzanne Kraszewski and the Solution Tree editorial team. Your continued support allows all preK–12 students to learn mathematics by experiencing daily lessons through the use of formative feedback and action processes.

We deeply appreciate the feedback given to us by our colleagues and reviewers. Thank you for your insights and ideas to strengthen the rubrics, tools, examples, and protocols in the book.

We are also grateful to our families. We are fortunate to have the support of loved ones as we do the work we so passionately pursue.

Finally, we are most grateful to you, the reader—the teacher and counselor, the mentor and coach, the team leader and school leader. Thank you for your daily effort to work together to ensure students learn mathematics as you help them to persevere, discover their own agency for learning, and engage in meaningful and vital routines for learning. It is your commitment to the belief that *every student can learn mathematics* that provides the hope for our students' future.

Solution Tree Press would like to thank the following reviewers:

Brian Buckhalter
Mathematics Coach and Consultant
Founder, Buck Wild About Math, LLC

Jason Cianfrance
Mathematics Teacher
Legacy High School
Broomfield, Colorado

Maria Everett
Mathematics Consultant
Bogue, North Carolina

Tracey Hulen
Mathematics Specialist
T.H. Educational Solutions
Fairfax, Virginia

Jennifer Rose Novak
Director of Curriculum, Instruction, and Assessment
Howard County Public Schools
Ellicott City, Maryland

Gina Rivera
Principal
Charter Oak International Academy
West Hartford, Connecticut

Sharon Rendon
Director of Professional Learning
CPM Educational Program

Gwendolyn Zimmermann
Author and Educational Consultant

Visit **go.SolutionTree.com/MathematicsatWork** to download the free reproducibles in this book.

Table of Contents

Reproducibles are in italics.

About the Authors

Mona Toncheff, an educational consultant and author, has over thirty years of experience in public education. She worked as both a mathematics teacher and a mathematics specialist for the Phoenix Union High School District in Arizona. In the latter role, she coached and provided professional development to high school teachers and administrators related to quality mathematics teaching and learning and effective collaborative teams. She currently serves as a supervisor teacher for the University of Arizona Teach Program.

Toncheff has supervised the culture change from teacher isolation to professional learning communities, created articulated standards and relevant district common assessments, and provided ongoing professional development on best practices, equity and access, technology, response to intervention, high-quality grading practices, and assessment for learning.

Toncheff has coauthored several Solution Tree and National Council of Supervisors of Mathematics (NCSM) publications. Her Solution Tree books include *Common Core Mathematics in a PLC at Work, High School* (2012); *Beyond the Common Core: A Handbook for Mathematics in a PLC at Work, High School* (2014); and *Activating the Vision: The Four Keys of Mathematics Leadership* (2016), and she is one of the lead coauthors of the *Every Student Can Learn Mathematics* series. Toncheff was the lead editor and contributing author for the NCSM books *Framework for Leadership in Mathematics Education* (2020), *Instructional Leadership in Mathematics Education* (2019), and *Culturally Relevant Leadership in Mathematics Education* (2022).

As a writer and consultant, Toncheff works with educators and leaders across the United States to build collaborative teams, empowering them with effective strategies for aligning curriculum, instruction, and assessment to ensure all students receive high-quality mathematics instruction.

A past president of NCSM, Toncheff served as an active board member of NCSM for over eleven years. In addition, she was a cofounding board member and past president of Arizona Mathematics Leaders. In 2009, she was named Phoenix Union High School District Teacher of the Year. In 2014, she received the Copper Apple Award for leadership in mathematics from the Arizona Association of Teachers of Mathematics, and in 2022, she received the AML Leadership Award from Arizona Mathematics Leaders.

Toncheff earned a bachelor of science from Arizona State University and a master of education in educational leadership from Northern Arizona University.

To learn more about Mona Toncheff's work, follow her @toncheff5 on Twitter.

Timothy D. Kanold, PhD, is an award-winning educator and author. He formerly served as director of mathematics and science and as superintendent of Adlai E. Stevenson High School District 125, a Model Professional Learning Community (PLC) district in Lincolnshire, Illinois.

Dr. Kanold has authored or coauthored and led more than thirty-seven textbooks and professional development books on K–12 mathematics, school culture, and school leadership, including his best-selling and 2018 Independent Publisher Book Award–winning book *HEART! Fully Forming Your Professional Life as a Teacher and Leader*. In 2021, he authored a sequel to *HEART!*, the book *SOUL! Fulfilling the Promise of Your Professional Life as a Teacher and Leader*, and most recently, he coauthored the best-selling book *Educator Wellness: A Guide for Sustaining Physical, Mental, Emotional, and Social Well-Being* (2022).

Dr. Kanold received the 2017 Ross Taylor / Glenn Gilbert National Leadership Award from the National Council of Supervisors of Mathematics, the international 2010 Damen Award for outstanding contributions to education from Loyola University Chicago, and the 1986 Presidential Award for Excellence in Mathematics and Science Teaching.

Dr. Kanold is committed to equity, excellence, and social justice reform for the improved learning of students and school faculty, staff, and administrators. He conducts inspirational professional development seminars worldwide with a focus on improving student learning outcomes in mathematics through a commitment to the PLC process. He also intentionally focuses on how to have a well-balanced, fully engaged professional life by practicing daily reflection and wellness routines.

Dr. Kanold earned a bachelor's degree in education and a master's degree in applied mathematics from Illinois State University. He received his doctorate in educational leadership and counseling psychology from Loyola University Chicago.

To learn more about Dr. Kanold's work, follow him @tkanold, #heartandsoul4ED, #MAWPLC, or #liveyourbestlife on Twitter.

Sarah Schuhl is an educational coach and consultant specializing in mathematics, professional learning communities, common formative and summative assessments, priority school improvement, and response to intervention (RTI). She has worked in schools as a secondary mathematics teacher, high school instructional coach, and K–12 mathematics specialist.

Schuhl was instrumental in the creation of a PLC in the Centennial School District in Oregon, helping teachers make large gains in student achievement. She earned the Centennial School District Triple C Award in 2012.

Schuhl grows learning in districts throughout the United States through large-group professional development and small-group coaching. Her work focuses on strengthening the teaching and learning of mathematics, having teachers learn from one another when working effectively as a collaborative team in a PLC, and striving to ensure the learning of each and every student through assessment practices and intervention. Her practical approach includes working with teachers and administrators to implement assessments for learning, analyze data, collectively respond to student learning, and map standards.

For Mathematics at Work, Schuhl coauthored *Engage in the Mathematical Practices: Strategies to Build Numeracy and Literacy With K–5 Learners* and (with Timothy D. Kanold) the *Every Student Can Learn Mathematics* series and the *Mathematics at Work™ Plan Book*, and was editor of the *Mathematics Unit Planning in a PLC at Work* series. Additionally, she contributed to the National Council of Supervisors of Mathematics (NCSM) publication *NCSM Essential Actions: Framework for Leadership in Mathematics Education*. On the subject of priority schools, Schuhl coauthored *School Improvement for All: A How-To Guide for Doing the Right Work* and *Acceleration for All: A How-To Guide for Overcoming Learning Gaps*. She contributed to *Charting the Course for Leaders: Lessons From Priority Schools in a PLC at Work* and *Charting the Course for Collaborative Teams: Lessons From Priority Schools in a PLC at Work*.

Previously, Schuhl served as a member and chair of the National Council of Teachers of Mathematics (NCTM) editorial panel for the journal *Mathematics Teacher* and secretary of NCSM. Her work with the

Oregon Department of Education includes designing mathematics assessment items, test specifications and blueprints, and rubrics for achievement-level descriptors. She has also contributed as a writer to a middle school mathematics series and an elementary mathematics intervention program.

Schuhl earned a bachelor of science in mathematics from Eastern Oregon University and a master of science in mathematics education from Portland State University.

To learn more about Sarah Schuhl's work, follow her @SSchuhl on Twitter.

Bill Barnes is the chief academic officer for the Howard County Public School System in Maryland. A past president of the Maryland Council of Teachers of Mathematics (MCTM), Barnes has also served nationally as a regional director and vice-president for NCSM: Leadership in Mathematics Education, and as the affiliate service committee Eastern Region 2 representative for the National Council of Teachers of Mathematics (NCTM).

Barnes is passionate about ensuring equity and access in mathematics for students, families, and staff. His experiences drive his advocacy efforts as he works to ensure opportunity and access for underserved and underperforming populations. He fosters partnerships among schools, families, and community resources in an effort to eliminate traditional educational barriers.

Barnes was the recipient of the 2003 Maryland Presidential Award for Excellence in Mathematics and Science Teaching. He was named Outstanding Middle School Math Teacher by the Maryland Council of Teachers of Mathematics and Maryland Public Television and Master Teacher of the Year by the National Teacher Training Institute.

Barnes earned a bachelor of science in mathematics from Towson University and a master of science in mathematics and science education from Johns Hopkins University and has served as an adjunct professor for Johns Hopkins University, the University of Maryland–Baltimore County, McDaniel College, and Towson University.

To learn more about Bill Barnes's work, follow him @ BillJBarnes on Twitter.

Jennifer Deinhart is an educational consultant and K–8 mathematics specialist. She is currently working as a mathematics instructional coach at Rose Hill Elementary, part of Fairfax County Public Schools in Falls Church, Virginia. During her time at Mason Crest Elementary in Annandale, Virginia, the school was recognized as the first Model PLC school to receive the DuFour Award. A passionate educator with more than twenty years of experience working with diverse populations within Title I schools, Deinhart leads collaborative teams of teachers to provide quality mathematics instruction.

As a consultant and presenter, Deinhart guides educators in effective teaming practices, such as the collection of quality evidence of student learning, progress monitoring that leads to intentional targeted re-engagement, and systematic processes for student self-assessment and goal setting. She joined the Mathematics at Work author team for *Mathematics Unit Planning in a PLC at Work, Grades PreK–2* and *Mathematics Unit Planning in a PLC at Work, Grades 3–5*. Deinhart has led multiple districts in adapting pacing guides and developing curriculum resources. She continues to support schools across the United States in deeply learning and implementing the PLC at Work process.

Deinhart received a bachelor's degree from Buffalo State University of the State University of New York and a master of education specializing in K–8 mathematics leadership from George Mason University.

Jessica Kanold-McIntyre is an educational consultant and author committed to supporting teacher implementation of rigorous mathematics curriculum and assessment practices blended with research-informed instructional practices. She works with teachers and schools around the United States to meet the needs of their students. Specifically, she specializes in building and supporting the collaborative teacher culture through the curriculum, assessment, and instruction cycle.

Kanold-McIntyre is currently the executive director of human relations and superintendent-elect in Aptakisic-Tripp Community Consolidated School

District 102 in Buffalo Grove, Illinois, where she has also served as a middle school principal, assistant principal, and mathematics teacher and leader. As principal of Aptakisic Junior High School, she supported her teachers in implementing initiatives, such as the Illinois Learning Standards; the Next Generation Science Standards; and the College, Career, and Civic Life Framework for Social Studies State Standards, while also supporting a one-to-one iPad environment for all students. She focused on teacher instruction through the PLC process, creation of learning opportunities around formative assessment practices, data analysis, and student engagement. She previously served as assistant principal at Aptakisic, where she led and supported special education, response to intervention (RTI), and English learner staff through the PLC process.

As a mathematics teacher and leader, Kanold-McIntyre strove to create equitable and rigorous learning opportunities for all students while also providing them engaging and challenging experiences to foster critical-thinking and problem-solving skills. As a mathematics leader, she developed and implemented a districtwide process for the Common Core State Standards in Illinois and led a collaborative process to create mathematics curriculum guides for K–8 mathematics, algebra 1, and algebra 2. She has served as a board member for the National Council of Supervisors of Mathematics (NCSM).

Kanold-McIntyre earned a bachelor of arts in elementary education from Wheaton College and a master of arts in educational leadership from Northern Illinois University. She will complete her doctorate in educational leadership at National Louis University.

To learn more about Jessica Kanold-McIntyre's work, follow her @jkanold on Twitter.

Matthew R. Larson, PhD, is an award-winning educator and author who served as the K–12 mathematics curriculum specialist for Lincoln Public Schools in Nebraska for more than twenty years. He served as president of the National Council of Teachers of Mathematics (NCTM) from 2016 to 2018. Dr. Larson has taught mathematics at the elementary through college levels and has held an honorary appointment as a visiting associate professor of mathematics education at Teachers College, Columbia University. He currently serves as the associate superintendent for instruction for Lincoln Public Schools, Nebraska.

Dr. Larson has coauthored several mathematics textbooks, professional books, and articles on mathematics education, and was a contributing writer on the influential publications *Principles to Actions: Ensuring Mathematical Success for All* (NCTM, 2014) and *Catalyzing Change in High School Mathematics: Initiating Critical Conversations* (NCTM, 2018). A frequent keynote speaker at national meetings, Dr. Larson is well known for his humorous presentations and their application of research findings to practice.

Dr. Larson earned a bachelor's degree and doctorate from the University of Nebraska–Lincoln.

To book Mona Toncheff, Timothy D. Kanold, Sarah Schuhl, Bill Barnes, Jennifer Deinhart, Jessica Kanold-McIntyre, or Matthew R. Larson for professional development, contact pd@SolutionTree.com.

Preface

By Timothy D. Kanold

In the early 1990s, I had the honor of working with Richard DuFour at Adlai E. Stevenson High School in Lincolnshire, Illinois. During that time, Rick—then principal of Stevenson—began his revolutionary work as one of the architects of the Professional Learning Communities at Work® (PLCs at Work) process. My role at Stevenson was to create, initiate, and incorporate the elements of the PLC process into the preK–12 mathematics programs, including the K–5 and 6–8 schools feeding into the Stevenson district.

In those early days of building the PLC process and collaborative culture, my colleagues and I exchanged knowledge about our personal growth and improvement as teachers and began to create and enhance student agency and self-efficacy for learning mathematics. As colleagues and team members, we taught, coached, and learned from one another. Our instructional design work, over time, became more transparent.

And yet, we did not know our mathematics *instruction* story. We did not have much clarity on an instructional vision that would significantly improve student learning and erase unintentional inequities caused by our private decision making for our mathematics assessment.

Through our work together, we realized that, without intending to, our isolated decisions about the teaching of mathematics often created the crushing consequence of poor student performance in a vertically connected curriculum like mathematics.

We also realized the benefit of belonging to something larger than ourselves. There is a benefit to learning about various teaching and assessing strategies from each other, *as professionals*. It was often in community that we found deeper meaning to our work, and strength in the journey, together.

This idea of a collaborative focus on the real work we do as mathematics teachers is at the heart of the *Every Student Can Learn Mathematics* series. The authors of this book believe that if teachers do the right work together and develop positive routines together, then just maybe every student can learn mathematics. This belief that began with the mighty team of mathematics teachers at Stevenson has been the driving force of the work of Mathematics in a PLC at Work for more than thirty years.

In this series, we emphasize the concept of *team actions*. We recognize some readers may be the only members of a grade level or mathematics course. In that case, we recommend you work with a colleague in a grade level or course above or below your own, or work with other job-alike teachers across a geographic region as technology allows. Your collaborative team becomes the engine that drives the PLC process forward.

A PLC in its truest form is "an ongoing process in which educators work collaboratively in recurring cycles of collective inquiry and action research to achieve better results for the students they serve" (DuFour, DuFour, Eaker, Many, & Mattos, 2016, p. 10). This book and the others in the *Every Student Can Learn Mathematics* series feature a wide range of voices, tools, and discussion protocols offering advice, tips, and knowledge for your PLC-based collaborative mathematics team.

The lead authors of the *Every Student Can Learn Mathematics* series—Sarah Schuhl and Mona Toncheff—have each been on their own journeys with the deep and collaborative PLC work for mathematics. They have spent significant time in the classroom as highly successful practitioners, leaders, and coaches of preK–12 mathematics teams, designing and leading the structures and the culture necessary for effective and collaborative team efforts. They, along with our additional coauthors and associates, have lived through and fiercely led the mathematics professional growth actions this book advocates within diverse preK–12 educational settings in rural, urban, and suburban schools.

In this book, we describe a highly successful mathematics instruction story. It is a story about daily decisions and choices regarding a balance of higher- and lower-level-cognitive-demand mathematical tasks to teach the essential learning standards of the unit. It is also about strategies and routines used for teaching those standards through a daily discourse balance during the mathematics lesson, and the most effective ways in which every mathematics lesson should begin and end—if our priority is to ensure student learning and engagement *during* the lesson. It is a preK–12 story that, when well implemented, will bring great satisfaction to your work as a mathematics professional.

For your grade-level or course-based collaborative team, helping students persevere and discover meaning during a mathematics lesson as *part of a formative process* for learning is where the most impactful work of your actual teaching and student learning and meaning is located. This formative process is at the heart of *Mathematics Instruction and Tasks in a PLC at Work*. We want to help your teacher team participate in meaningful discussions about how to prepare and then execute a mathematics lesson design that will significantly increase student learning and effort each day.

We hope you will join us on this journey of significantly improving student learning in mathematics by leading and improving your mathematics instruction story for your team, your school, or your district. The conditions and the actions for adult learning of mathematics together reside in the pages of this book.

We hope the personal stories we tell, the tools we provide, and the reflection opportunities we include serve your daily professional work well as you, your colleagues, and your collaborative team pursue student learning in a discipline we all love—mathematics.

Introduction

If you are a teacher of mathematics, then this book is for you! Whether you are a novice or a master teacher; an elementary, middle, or high school teacher; or a rural, suburban, or urban teacher, this book is for you. It is for all teachers and support professionals who are part of the preK–12 mathematics learning experience.

Teaching mathematics so *each and every student* learns the preK–12 mathematics curriculum, develops a positive mathematics identity, and becomes empowered by mathematics is a complex and challenging task. Your professional life as a mathematics teacher is not easy.

Yet when you work *together*, as an effective collaborative team, throughout a mathematics unit of study, you diminish potential levels of inequity. These rigor inequities surface in your choice of mathematical activities and tasks for student learning before, during, and after the lesson. A lack of rigor consistency in what you expect students to know and be able to do, how you will know when students have learned, what you will do *during the lesson* when students have not learned, and how you will proceed as your students are demonstrating learning creates a wide variance that, left unchecked, causes deep inequities in mathematical understanding as your students pass from grade to grade and course to course.

Equity and PLCs

The PLC at Work process is one of the best and most promising models your school or district can use to build a more equitable response for student learning. Richard DuFour, Robert Eaker, and Rebecca DuFour, the architects of the PLC process, designed the process around three big ideas and four critical questions that place learning, collaboration, and results at the forefront of our work as teachers (DuFour et al., 2016). As DuFour and colleagues (2016) explain in their large cadre of work, schools and districts that commit to the PLC transformation process rally around the following three big ideas.

1. **A focus on learning:** Teachers focus on learning as the fundamental purpose of the school rather than on teaching as the fundamental purpose.
2. **A collaborative culture:** Teachers work together in teams to interdependently achieve a common goal or goals for which members are mutually accountable.
3. **A results orientation:** Team members are constantly seeking evidence of the results they desire—high levels of student learning.

Additionally, teacher teams within a PLC focus on four critical questions (DuFour et al., 2016) as part of their instruction and task-creation routines used to inspire student learning:

1. What do we want all students to know and be able to do *in class*?
2. How will we know if students learn it *in class*?
3. How will we respond *in class* when some students do not learn?
4. How will we extend the learning *in class* for students who are already proficient?

The four critical questions of a PLC provide an equity lens for your professional work during instruction. Notice the intentional adaptation of the critical questions around the words *in class*. This is intentional, as this book is about the student learning process during the lesson and the potential gaps that will exist if you and your colleagues do not agree on the rigor for the mathematical tasks you use to answer the question, "What do we want all students to know and be able to do in class today?"

For you and your colleagues to answer these four PLC critical questions well during the lesson requires the development, use, and understanding of lesson-design criteria that will cause students to engage in the lesson, persevere through the lesson, and embrace their errors as they demonstrate learning pathways for the various mathematics tasks you present to them.

The concept of your team reflecting together and then taking action around the right mathematics lesson-design work is a point of emphasis in the *Every Student Can Learn Mathematics* series. Your team's reflection will increase students' agency (their voice, ownership, perseverance, and action during learning) as you work together to create a unified, robust formative process for helping students own their responses during class when they are and are not learning.

The Reflect, Refine, and Act Cycle

Figure I.1 illustrates the reflect, refine, and act cycle, our perspective about the process of lifelong learning—for you and for your students. The very nature of the education profession is about the development of skills toward learning. Those skills are part of an ongoing process you pursue with your colleagues.

More importantly, the reflect, refine, and act cycle is a *formative* student learning cycle described throughout all books in the *Every Student Can Learn Mathematics* series. When your teacher team embraces mathematics learning as a *process*, your students reflect, refine, and act by asking the following questions.

- **Reflect:** "How well do I make sense of the mathematics task, solve it with a chosen strategy, and determine, 'Is this the best solution strategy?'"
- **Refine:** "How well do I learn mathematics based on feedback about my work, solution pathways, and any possible errors?"
- **Act:** "How well do I persevere, apply learning from the mathematics task to future tasks, and determine what I have learned that I can use again?"

The intent of this *Every Student Can Learn Mathematics* series is to provide you with a systemic way to structure and facilitate deep team discussions necessary to lead an effective and ongoing adult and student learning process each and every school year.

Mathematics in a PLC at Work Framework

The *Every Student Can Learn Mathematics* series includes three books that focus on a total of five teacher team actions within three larger categories.

1. *Mathematics Assessment and Intervention in a PLC at Work, Second Edition*
2. *Mathematics Instruction and Tasks in a PLC at Work, Second Edition*
3. *Mathematics Homework and Grading in a PLC at Work*

Figure I.2 shows each of these three categories and the teacher team actions within each. These five team actions focus on the nature of the ongoing, unit-by-unit professional work of your teacher team and how you should respond to the four critical questions of a PLC at Work (DuFour et al., 2016).

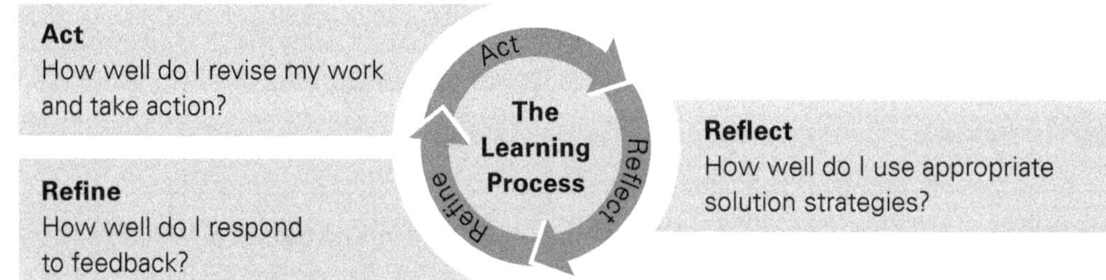

Figure I.1: Reflect, refine, and act cycle for formative student learning.

Every Student Can Learn Mathematics series' Team Actions Serving the Four Critical Questions of a PLC at Work	1. What do we want all students to know and be able to do?	2. How will we know if students learn it?	3. How will we respond when some students do not learn?	4. How will we extend the learning for students who are already proficient?
Mathematics Assessment and Intervention in a PLC at Work, Second Edition				
Team action 1: Develop high-quality common assessments for the agreed-on essential learning standards.	■	■		
Team action 2: Analyze and use common assessments for formative student learning and intervention.			■	■
Mathematics Instruction and Tasks in a PLC at Work, Second Edition				
Team action 3: Develop high-quality mathematics lessons for daily instruction.	■	■		
Team action 4: Analyze and use effective lesson-design elements to provide formative feedback and build student perseverance.			■	■
Mathematics Homework and Grading in a PLC at Work				
Team action 5: Develop and use high-quality common grading components and formative grading routines.	■	■	■	■

Figure I.2: Mathematics in a PLC at Work framework.

Visit ***go.SolutionTree.com/MathematicsatWork*** *for a free reproducible version of this figure.*

Most commonly, a collaborative team consists of two or more teachers who teach the same grade level or course. Through your focused work addressing the four critical questions of a PLC (DuFour et al., 2016), you provide every student in your grade level or course with equitable learning experiences and expectations, opportunities for sustained perseverance, and robust formative feedback, regardless of the teacher they receive.

If, however, you are a *singleton* (a teacher who does not have a colleague who teaches the same grade level or course), you will have to determine who it makes the most sense for you to work with as you strengthen your lesson design and student feedback skills. Leadership consultant and author Aaron Hansen (2015) and consultants and authors Brig Leane and Jon Yost (2022) suggest the following possibilities for creating teams for singletons.

- Vertical teams (for example, a primary school team of grades K–2 teachers or a middle school mathematics department team for grades 6–8)
- Virtual teams (for example, a team comprising teachers from different sites who teach the same grade level or course and collaborate virtually with one another across geographic regions)
- Grade-level or course-based teams (for example, a team where teachers expand to teach and share two or three grade levels or courses instead of each only teaching all sections of one grade level or course)

About This Book

In this book, you will find spaces to write reflections about your practice. You are also provided with team discussion protocols and tools to make your team meeting discussions focused, mindful, and meaningful.

The team discussion tools and protocols are designed for you to feel confident and comfortable in discussing your lesson content and process with one another, and in moving toward greater transparency in your instructional practice and understanding of the standards with colleagues. In this book, you will also find personal stories from the authors' experiences that shed light on the impact your team actions have on classroom practice.

The Mathematics at Work™ instructional framework appears in a special section following this introduction (see page 5). You and your teacher team will reflect on the six research-affirmed lesson-design elements for the high-quality instruction that is the focus of this book.

Next, chapters 1–6 will explore in more detail each of the research-affirmed lesson-design elements. You and your teacher team will learn how to implement those specific lesson-design elements to provide formative feedback and build student perseverance.

Chapters 7 and 8 address high-quality Tier 1 interventions and team instructional learning experiences through lesson studies and instructional rounds. You will explore your team's responsibility to continually evaluate the effectiveness of your core instruction and adapt instruction as needed to meet the needs of each and every student. You will examine strategies for how to analyze student interactions during your lessons, how to make your instructional routines more transparent, and how to evaluate the impact of your core instructional decisions.

You can visit **go.SolutionTree.com/MathematicsatWork** to access the free reproducibles that appear in this book. (See the QR code printed throughout this book for easy access to the materials.) In addition, online you will find grade-level or course-based lesson-design samples along with a comprehensive list of free online resources—"Online Resources Reference Guide for Mathematics Support"—to support your work in mathematics teaching and learning.

As you embrace the belief that you and your colleagues together can overcome the many obstacles you face each day in the work of your PLC, then *every student can learn mathematics* just may become a reality in your school.

The Mathematics at Work™ Lesson-Design Framework

You and your teaching colleagues have a lot of power—power that resides in the lesson-design decisions you make every day. In some ways, Martin Luther King Jr. provides a description central to the essence of great mathematics instruction with his quote, "The function of education is to teach one to think intensively and to think critically" (as cited in Obeng, 2022).

Admittedly, in the 21st century education environment, teachers use the word *persevere* instead of *think intensively* and the phrase *higher-level-cognitive-demand reasoning* instead of *think critically*. And in every lesson design and instructional moment, your students should do these actions with *balance*. As you and your colleagues prepare mathematics lessons each day, and for each unit of study, your lesson-design choices shape—and can make or break—students' learning experiences. There are two specific team actions (from figure I.2, page 3) that support your team's daily lesson-design decisions.

- **Team action 3:** Develop high-quality mathematics lessons for daily instruction.
- **Team action 4:** Analyze and use effective lesson-design elements to provide formative feedback and build student perseverance.

Team action 3 is about you and your team preparing each of the six lesson-design elements in advance of the lesson to significantly improve student learning. In some cases, effective lesson design, coupled with effective lesson implementation, can double the speed of learning toward the essential mathematics standards of the unit (Popham, 2011). Team action 3 supports the *what* of lesson design, and team action 4 supports the *how* of lesson design.

The following six elements of effective mathematics lesson design are the focus of each of the first six chapters in this book.

1. Essential learning standards: the *why* of the lesson
2. Prior-knowledge routines
3. Mathematical language routines
4. Lower- and higher-level-cognitive-demand mathematical task balance
5. Mathematical discourse routines
6. Lesson-closure routines

Each chapter explores one of the lesson-design elements and how you and your teacher team intentionally design instructional routines. Each chapter also focuses on how to implement each element to provide formative feedback and foster student perseverance during the lesson. Team action 4 ensures your collaborative team reaches clarity on how to effectively respond in class to the third and fourth critical questions of a PLC (DuFour et al., 2016).

3. How will you engage your students in the essential learning standard, and how will you collect evidence of learning during the lesson?

4. How will you vary the types of activities and tasks to ensure you extend student engagement and promote perseverance throughout the lesson? And, most importantly, how will you fully engage the students during the discourse of the lesson?

Each chapter provides you with discussion tools, samples, and insights that will help you answer these questions for every mathematics lesson, every day.

Think about the lessons you taught during your most recent unit of mathematics for your grade level or course. Reflect on the most essential standards you were trying to help students learn. Use the seven questions in figure I.3 to focus your team's lesson-design discussions.

The purpose of any mathematics lesson is about facilitating your students' learning of the essential standard or learning target for that day. The choices you make regarding the mathematics problems and tasks are the means for how you get your students to the final destination: *learning the essential standards for the unit.* How you will use those mathematical task choices in class becomes one of your greatest professional responsibilities.

Another equity-based mathematics teaching practice offered by Julia Aguirre, Karen Mayfield-Ingram, and Danny Bernard Martin (2013) is going deep with mathematics. They specifically argue that mathematics lessons "include high cognitive demand tasks that support and strengthen students' development of the strands of mathematical proficiency" (Aguirre et al., 2013, p. 43). In this series, *cognitive demand* is described as the level of and complexity of the reasoning required by the students for each mathematical task during the lesson. Students learn best when they are engaged in mathematical tasks that provide opportunities to make sense of the meaningful mathematics and to engage in productive perseverance. Cognitive demand is explored more in chapter 4 (page 45).

Thus, a mathematics lesson is about so much more than just "doing" a bunch of mathematics problems or tasks with your students. The choices of the tasks you use to teach the lesson are very important. In fact, they give you a lot of power as a teacher, but they are not in and of themselves the reason for or the purpose of the mathematics lesson.

The ultimate purpose of any mathematics lesson is to *maximize student engagement, communication, and perseverance during the lesson based on the tasks you have chosen* in order to help students learn the essential learning standards for each unit of mathematics.

You and your team can use figure I.4 (page 8) to evaluate the quality of your current mathematics lessons. It will help you identify areas of strength and areas for lesson-design growth. (Visit **go.SolutionTree.com/MathematicsatWork** or scan the QR code that appears at the end of this chapter to download this free reproducible tool along with additional support tools and sample documents.)

Before you dive into the details of each element, take a moment to examine your current use of the six mathematics lesson-design elements listed in the teacher reflection.

TEACHER *Reflection* 

How often do you currently implement the six lesson-design elements? Rank yourself and your colleagues on a scale of 1 to 10 for each criterion (1 = we don't use it at all; 5 = we implement it sometimes, and it could use improvement; and 10 = we are awesome at it and use it all the time).

1. Essential learning standards: the *why* of the lesson
2. Prior-knowledge routines
3. Mathematical language routines as part of instruction
4. Lower- and higher-level-cognitive-demand mathematical task balance
5. Whole-group and small-group discourse balance
6. Lesson-closure routines for student reflection

Directions: Use the following prompts to guide discussion about your current lesson design.

Purpose of the lesson:

1. What is the fundamental purpose of a mathematics lesson?

2. How do you inform your students of the relevance—the *why*—for the day? Do you inform them in writing or verbally?

Essential elements of a mathematics lesson: Respond to each question with a yes or no, and then briefly explain how you make the lesson-design choice or why you do not make the lesson-design choice.

3. Do you choose a warm-up or prior-knowledge mathematics question or task to begin each lesson?

4. Do you choose to discuss and connect key vocabulary words for the lesson?

5. Do you choose lower- and higher-level-cognitive-demand mathematical tasks that align to the essential standard for each lesson? If so, is it a collaborative teacher team activity?

6. Do you intentionally choose whole-group *and* small-group discourse activities as part of the lesson experience?

7. Do you close each lesson with a student-led summary?

Figure I.3: Teacher team discussion tool—Collaborative lesson-design elements.

Visit ***go.SolutionTree.com/MathematicsatWork*** *for a free reproducible version of this figure.*

High-Quality Lesson-Design Elements	Description of Level 1	Requirements of the Indicator Are Not Present	Limited Requirements of the Indicator Are Present	Substantially Meets the Requirements of the Indicator	Fully Achieves the Requirements of the Indicator	Description of Level 4
1. Essential learning standards: the *why* of the lesson	The lesson references an essential learning standard but doesn't have a clear learning target, and there is no evidence of consistent learning standard or target language across the collaborative team.	1	2	3	4	The lesson design declares a daily learning target aligned to an essential learning standard for the unit. Teachers share a context for that learning target with students during the lesson and clearly relate the story arc for *why* the standard is important mathematics to learn.
2. Prior-knowledge routines	Either there is no prior-knowledge routine to the lesson content or the routine exists but does not clearly support students' accessing prior knowledge needed for the lesson.	1	2	3	4	There is a prior-knowledge routine that includes an opportunity for students to work together and engage in prior-knowledge thinking about the mathematics necessary to prepare them to persevere during the lesson.
3. Mathematical language routines	The lesson does not address mathematical language (vocabulary and notations) explicitly with a formal plan for ensuring student demonstrations with clarity.	1	2	3	4	There is evidence of focused mathematical language routines designed to support student meaning making, sense making, and communication of the mathematics content across all grade-level or course-based teams.
4. Lower- and higher-level-cognitive-demand mathematical task balance	There is no evidence of an intentional and balanced use of lower- and higher-level-cognitive-demand mathematical tasks. There are no specific strategies for engaging students in the sense making or application of the content.	1	2	3	4	There is an intentional choice made to balance higher- and lower-level-cognitive-demand mathematical tasks within the lesson design. There is a specific focus on formative feedback and action routines with peers and the teacher during the lesson.
5. Whole-group and small-group mathematical discourse routines	There are no specific strategies for how students will discuss and share their thinking with their peers. The lesson plan relies solely on whole-group, teacher-led discourse. During the lesson, the teacher asks nearly all the questions and is the only one evaluating the student responses.	1	2	3	4	There are intentional plans for the type of discourse (whole group or small group) that students will experience for each mathematical task and portion of the mathematics lesson. There is a commitment to balancing student time to process and communicate with one another (what you see and hear the students doing) against the time given to teacher-directed instruction.
6. Lesson-closure routines	The lesson does not include a summary, or the summary is teacher centered (as opposed to student centered). There is no opportunity for students to self-evaluate to determine if they meet and understand the learning target for the day.	1	2	3	4	The lesson includes a student-led closure activity to self-assess if the lesson helped students understand the learning target or essential learning standard for the day. During implementation of the lesson-closure routines, teachers use the responses to plan next steps.

Figure I.4: Mathematics in a PLC at Work instructional framework and lesson-design evaluation rubric.

Visit ***go.SolutionTree.com/MathematicsatWork*** *for a free reproducible version of this figure.*

Effective mathematics instruction rests in part on your careful lesson design (Morris, Hiebert, & Spitzer, 2009). Your collaborative team is uniquely structured to provide the time and support it needs to interpret the standards, embed student-engaged discourse routines into daily lessons, and reflect together on the effectiveness of your implementation, sharing evidence of student learning.

You can use the design tool during your planning for the instruction of the mathematics lesson as a way to support the focus and design of the mathematics tasks you choose to teach each learning standard.

The appendix (page 111) presents the Mathematics in a PLC at Work lesson-design tool that organizes the six lesson-design elements and helps ensure your teacher team reaches mathematics lesson clarity on all four of the PLC critical questions.

1. What do we want all students to know and be able to do? (The learning targets)
2. How will we know if students learn it? (The assessment instrument and tasks)
3. How will we respond when some students do not learn? (Using a formative assessment process)
4. How will we extend the learning for students who are already proficient? (Using a formative assessment process)

Your collaborative team is most likely using some type of lesson-planning format or tool currently. What separates the mathematics lesson-design tool in this book from most other lesson-planning models is the focus on directly planning tasks from the student perspective in order to support students in reaching proficiency on the mathematics content standards of the lesson.

In addition to the six lesson-design elements, there are a few additional planning components that will contribute positively to your lesson-design process.

First, under the essential learning standard, there is a place to list the essential *content* learning standard. However, there is also a place to list the *process* learning standard. That references how you hope your students process learning the content standard during the lesson. Reference the National Council of Teachers of Mathematics (NCTM) process standards (see www.nctm.org/Standards-and-Positions/Principles-and-Standards/Process), your specific state or province's mathematical practice standards, or the mathematical proficiencies in the National Research Council's book *Adding It Up* (Kilpatrick, Swafford, & Findell, 2001).

Second, the Mathematics in a PLC at Work lesson-design tool also specifically asks you and your team to consider both student and teacher actions during lesson planning. It is designed to help you see learning through the eyes of students and help them become owners of their learning.

Third, as we describe in more detail, the Mathematics in a PLC at Work lesson-design tool will help you plan for formative assessment feedback questions and student evidence of learning during the mathematics lesson. These communication choices include:

- How do you and your colleagues expect students to express their ideas, questions, insights, and difficulties?
- Where, when, and between whom should the most significant conversations be taking place (student to teacher, student to student, or teacher to student)?
- How do you use questions to facilitate small-group peer-to-peer discourse and differentiate learning around higher-level-cognitive-demand tasks during the lesson?
- How approachable and encouraging should you be as students explore? Do students see you and other students as reliable and valuable learning resources?

Throughout the book, you will also be asked to reflect on and refine your lesson-design and lesson-implementation routines in order to achieve such a lofty expectation. You will reflect on how your current lessons become more coherent by meeting the demands of the four critical questions of a PLC (DuFour et al., 2016).

You might be surprised, but there is a theme that runs through mathematics instruction and lesson design when working as part of a collaborative mathematics team within a PLC at Work culture. Ready?

It's *balance and perseverance.*

Your daily designing and planning for a mathematics lesson can easily fall into a routine that is unbalanced in its mathematical task selection, strategies used to teach the lesson, and student discourse and engagement during the lesson. Without ongoing team discussion with your colleagues about your daily lesson design, you can unintentionally cause deep inequities in student learning.

- Do you know the following daily lesson routines of your colleagues?
- Do you each declare the mathematics learning target to be learned or the essential questions to be answered daily?
- Do you each connect every mathematics task used during the lesson to the standard for the day?
- Do you each balance the use of lower-level-cognitive-demand tasks (procedural knowledge with rote routines) with the use of higher-level, open-ended mathematical tasks? Do you each use application and mathematical modeling tasks during the unit?
- Do you each teach the academic vocabulary of the daily lesson?
- Do you each use a formative learning process that actively engages students during the lesson?
- Do you each use technology or other mathematical models as a routine part of the lesson design?

Wide variances in your daily decision making can cause a rigor inequity for students in the same grade level or course. In a vertically connected curriculum like mathematics, this variance can cause learning gaps as students progress through the grades.

Significant lesson-design differences may exist with how the lesson begins and ends as well. You may use prior-knowledge routines every day with a student-led closure activity. However, your colleagues may not.

Mathematics lessons have a lot of daily choices. And those choices should be designed to help your students demonstrate "productive perseverance" during a mathematics lesson and persevere through the variety of mathematics tasks they must do to demonstrate their learning (M. Larson, personal communication, July 30, 2017).

In this book, there is intentional guidance to help you and your colleagues reflect on your current lesson-design elements, compare your current instructional routines against high-quality standards of mathematics lesson design, and then develop and use lessons that effectively engage students with those lesson elements.

The benefit of these six lesson-design elements will be improved student perseverance in class that results in greater memory and retained learning of the expected mathematics standards for your grade level or course.

Visit **go.SolutionTree.com/MathematicsatWork** for free reproducible versions of tools and protocols that appear in this book, as well as additional online only materials.

CHAPTER 1

Essential Learning Standards: The *Why* of the Lesson

When teachers refer to the goals of the lesson during instruction, students become more focused and better able to perform self-assessment and monitor their own learning.

—*Shirley Clarke, Helen Timperley, and John Hattie*

Perhaps one of the most glaring weaknesses of many mathematics lessons is the lack of student facility to state with clarity, understanding, and precision the actual learning standard for the lesson. If you were to ask most students as they move on to the next activity (elementary school) or walk out of class (middle school or high school) the question, "What did you learn in math class today?" they most likely would respond, "I learned how to do some math problems."

For many students, learning mathematics is about doing a bunch of disjointed and seemingly meaningless mathematical tasks, mathematics problems, mathematics questions, or just "math stuff." Most students will not be able to, when called on, explain the reason they did the mathematics tasks their teacher chose for the lesson that day. They are unable to explain the *why* of the lesson or describe the standard for the lesson.

For example, in a fifth-grade classroom, the intended learning standard might be, *I can understand the result when dividing a fraction by another fraction less than or greater than 1.* Yet when asked what they learned today in class, at best, students might say, "We did math problems with fractions."

Pushed further, if students are asked, "Why did you learn this today? What was the purpose of the mathematics lesson?" or "How is what you learned today connected to what you are learning in the unit or to the content you have already learned?" they generally have no clear response to the question. Connecting the story arcs of the learning standards is generally not part of mathematics lesson design. Thus, knowing the relevance of the lesson and understanding and providing clarity to the mathematics standard are the first elements of lesson design for you and your team to discuss.

NCTM's groundbreaking books *Principles to Actions* (2014) and the *Catalyzing Change* series (2018, 2020a, 2020b) establish eight research-informed daily instructional strategies and routines. NCTM indicates a mathematics teaching priority action should be to "establish mathematics goals to focus learning." This expectation requires more than merely identifying mathematical activities for the lesson. The essential learning standard communicates the expectation of the learning and makes the outcomes of the lesson less ambiguous for both the students and the teacher team (Kirschner, Hendrick, & Heal, 2022).

Tim Kanold's personal story on page 12 illustrates the importance of gaining team clarity about what students are learning and why understanding the essential learning standards is the first of the six essential design elements you and your team will tackle—partially because of the immediate effects on student perseverance and effort and partially because of the forced teacher clarity for understanding the standards.

Personal Story  TIMOTHY KANOLD

When I first got to Stevenson, I was struck by our general lack of knowledge about why we were teaching certain standards each day. It was as if the lessons just fell from the sky with no rhyme or reason. As teachers of mathematics, we very rarely expected our students to know why the lesson was so essential—and I am not sure we always had great clarity on why, either. We soon decided that every mathematics lesson we taught would include, in writing, the context of the lesson: *what happened before this lesson, and what will happen after this lesson.*

We noticed an immediate increase in student perseverance and effort during the lesson. We made sure to connect each mathematics task we chose for the lesson directly to the essential learning standard of the lesson; we made a commitment to tell our students the *why* every day. If we were not clear about the standard and understanding the *why*, we asked each other or did some fact finding. Without knowing it, we were into progressions long before the idea of mathematics learning progressions as part of an ongoing story became popular!

Essential Standards Routines

As you begin to design lessons with the essential learning standards in mind (team action 3), you and your colleagues should develop clarity on the essential standards for each mathematics unit. This requires your teacher team to break down the two to five essential learning standards for each unit into the daily learning targets for your lessons with an eye toward choosing and aligning the tasks for each lesson.

Essential Learning Standards

For the purposes of this book, the essential learning standards represent the two to five content clusters that will ultimately be assessed for student learning and progress during and at the end of a mathematics unit. These standards are part of your students' assessment analysis for grading purposes described in *Mathematics Assessment and Intervention in a PLC at Work, Second Edition* (Schuhl, Kanold, Toncheff, et al., 2024) for the *Every Student Can Learn Mathematics* series and the *Mathematics Unit Planning in a PLC at Work* series. You use the essential learning standards to assess student understanding on all common mid- and end-of-unit assessments. These clustered mathematics standards act as the driver for organizing your unit assessments and providing student intervention and support that take place during and after a unit of study. This is why identifying the essential learning standards for the mathematics unit is so important.

However, from a daily lesson-planning perspective, it becomes necessary to unpack those essential learning standards into daily learning targets. Your team "unpacks" the standard by breaking it down into *daily learning targets* that reflect the conceptual knowledge, proficiency skills, and applications necessary for a student to demonstrate understanding of each essential learning standard for the unit. In some cases, an essential mathematics standard may carry over from one unit to the next.

Fortunately, the lesson-design elements in this book provide a sequential way to think about the elements of your lesson design. These elements serve to help you and your team members further unpack each essential learning standard for the mathematics unit as you consider the academic vocabulary, the cognitive demand of the chosen tasks, and the specific purpose for each daily learning target.

Learning Targets

Learning targets, then, are your daily learning objectives for the lesson. Often there may be several learning targets and mathematics lessons to help you achieve an essential learning standard for the unit. This is why establishing mathematics targets to focus learning "sits at the top" of a framework of research-informed instructional strategies, because it signifies that "setting goals [outcomes] is the starting point for all instructional decisions" (Smith, Steele, & Raith, 2017, p. 195).

These targets can be written in student-friendly language (*I can* statements).

Think of daily learning targets as a subset for each of the two to five essential learning standards for the unit. To help you compare and contrast essential learning standards and daily learning targets, figures 1.1–1.4 (pages 14–17) provide sample illustrations of a preK unit on composing and decomposing numbers up to 5, a second-grade base-ten unit, a seventh-grade statistics unit, and a high school algebra 1 linear equations unit. The essential learning standards in the left column are more formal descriptions, similar to how standards may appear in your state (or provincial) standards document. In the middle column, the essential learning standard is rewritten as it would appear on the mathematics unit assessments or tests in student-friendly *I can* language. The right column highlights an unwrapping of each standard into daily learning targets you might use for your daily lesson design and planning. It uses a combination of verbs and noun phrases from the original formal essential learning standard.

Your team should establish the connection between the essential learning standards for the unit and the mathematics tasks or problems you choose to teach those standards each day. What is the context for the progression of the standards in the unit? Does each daily learning target reveal a clear purpose as part of a bigger picture of learning? And, moreover, how will you explain and communicate this purpose to the students with clarity?

After you read the sample most appropriate to your grade level, answer the questions in the teacher reflection next.

TEACHER *Reflection*

How does your team create, develop, and communicate your understanding of the essential learning standards for the unit?

How do you break down the essential learning standards for the unit into daily learning targets for your lesson-design purposes?

Grade PreK Composing and Decomposing Numbers Up to 5 Unit: Essential Standards		
Formal Unit Standards (States standard language)	**Essential Learning Standards for Assessment and Reflection** (Uses student-friendly language)	**Daily Learning Targets** (Explains what students have to know and be able to do; unwrapped standards)
1. Given a set containing 5 or fewer concrete objects, create a set of objects that has the same as, 1 more than, or 1 less than the given set.	I can show the same number as you. I can show more or less than you.	• Count up to 5 objects and say how many. • Create a set of objects that is the same as the number of objects displayed (up to 5 objects). • Create a set of objects that is 1 more or 1 less than the number of objects displayed (up to 5 objects).
2. Model and solve problems within a familiar context with sums and differences within 5, using concrete objects.	I can show a math story with cubes. (Insert other objects as they are used.)	• Listen to familiar story contexts involving teacher modeling with concrete and real-life objects, such as, "I have 2 cookies. My teacher gives me 1 more cookie. How many cookies do I have?" • Model familiar story contexts with concrete and real-life objects, such as, "I have 4 crackers. I eat 1 cracker. How many crackers do I still have?"
3. Demonstrate the part-whole relationships for numbers up to 5 using concrete objects.	I can show a number on a five frame. I can show a number using my fingers.	• Represent numbers with counters on a five frame using one color and a combination of two colors. • Represent numbers up to 5 using fingers on one hand and a combination of fingers on both hands.

Figure 1.1: Sample essential learning standards for a preK composing and decomposing numbers up to 5 unit.

Grade 2 Apply and Extend Base-Ten Understanding Unit: Essential Learning Standards		
Formal Unit Standards (States standard language)	**Essential Learning Standards for Assessment and Reflection** (Uses student-friendly language)	**Daily Learning Targets** (Explains what students have to know and be able to do; unwrapped standards)
1. Understand that the three digits of a three-digit number represent amounts of hundreds, tens, and ones; for example, 706 equals 7 hundreds, 0 tens, and 6 ones. Understand the following as special cases: • The number 100 can be thought of as a bundle of ten tens—called a hundred. • The numbers 100, 200, 300, 400, 500, 600, 700, 800, and 900 refer to one, two, three, four, five, six, seven, eight, or nine hundreds (and 0 tens and 0 ones).	I can understand the meaning of the hundreds, tens, and ones in a three-digit number.	• Understand the three digits of a number represent amounts of hundreds, tens, and ones. • Know how to make a hundred using groups of ten. • Demonstrate the value or meaning of 100, 200, 300, 400, 500, 600, 700, 800, and 900 using groups of one hundred.
2. Count within 1,000; skip count by fives, tens, and hundreds.	I can count within 1,000.	• Count within 1,000. • Skip count by fives within 1,000. • Skip count by tens within 1,000. • Skip count by hundreds within 1,000.
3. Read and write numbers to 1,000 using base-ten numerals, number names, and expanded form.	I can read and write numbers using base-ten numerals, number names, and expanded form.	• Read and recognize numbers to 1,000s. • Write numbers to 1,000. • Understand the numerals, number names, and expanded form of numbers to 1,000.
4. Compare two three-digit numbers based on meanings of the hundreds, tens, and ones digits, using >, =, and < symbols to record the results of comparisons.	I can compare two numbers, explain my thinking, and write my answer using >, =, or < symbols.	• Recognize and use the >, =, < symbols to compare numbers. • Recognize the value of a three-digit number using place value understanding.

Figure 1.2: Sample essential learning standards for a grade 2 applying and extending base-ten understanding unit.

Grade 7 Statistics Unit: Essential Learning Standards		
Formal Unit Standards (States standard language)	**Essential Learning Standards for Assessment and Reflection** (Uses student-friendly language)	**Daily Learning Targets** (Explains what students have to know and be able to do; unwrapped standards)
1. Understand that the probability of a chance event is a number between 0 and 1 that expresses the likelihood of the event occurring. Larger numbers indicate greater likelihood. A probability near 0 indicates an unlikely event, a probability around $\frac{1}{2}$ indicates an event that is neither unlikely nor likely, and a probability near 1 indicates a likely event.	I can explain the meaning of a probability for a chance event.	• Develop a general understanding of the likelihood of events occurring by realizing that probabilities fall between 0 and 1. • Use the terms *likely* and *unlikely* to describe the probability fractions represent. • Understand that a probability near 0 indicates an unlikely event, a probability around $\frac{1}{2}$ indicates an event that is neither unlikely nor likely, and a probability near 1 indicates a likely event. • Understand what the sum of the probability of an event happening and not happening is.
2. Approximate the probability of a chance event by collecting data on the chance process that produces it and observing its long-run relative frequency, and predict the approximate relative frequency given the probability.	I can approximate the probability of a chance event using relative frequency from data collected in a chance process.	• Define a chance event and gather data on the chance process. • Demonstrate understanding of a relative frequency and predict the relative frequency given the probability.
3. Develop a probability model and use it to find probabilities of events. Compare probabilities from a model to observed frequencies; if the agreement is not good, explain possible sources of the discrepancy. ‣ Develop a uniform probability model by assigning equal probability to all outcomes, and use the model to determine probabilities of events. ‣ Develop a probability model (which may not be uniform) by observing frequencies in data generated from a chance process.	I can develop, use, and evaluate probability models.	• Assign equal probability to all outcomes to develop a uniform probability model, and use the model to determine probabilities of events. • Observe frequencies in data generated from a chance process to develop a probability model (which may not be uniform).
4. Find probabilities of compound events using organized lists, tables, tree diagrams, and simulation. ‣ Understand that, just as with simple events, the probability of a compound event is the fraction of outcomes in the sample space for which the compound event occurs. ‣ Represent sample spaces for compound events using methods such as organized lists, tables, and tree diagrams. For an event described in everyday language, identify the outcomes that compose the event in the sample space. ‣ Design and use a simulation to generate frequencies for compound events.	I can find probabilities of compound events using organized lists, tables, tree diagrams, and simulation.	• Understand that, just as with simple events, the probability of a compound event is the fraction of outcomes in the sample space for which the compound event occurs. • Develop probability models to find the probability of simple and compound events (organized lists, tables, tree diagrams, and simulations). • Represent sample spaces for compound events. • Design and use a simulation to generate frequencies for compound events.

Figure 1.3: Sample essential learning standards for a grade 7 statistics unit.

Algebra 1 Linear Equations Unit: Essential Learning Standards		
Formal Unit Standards (States standard language)	**Essential Learning Standards for Assessment and Reflection** (Uses student-friendly language)	**Daily Learning Targets** (Explains what students have to know and be able to do; unwrapped standards)
1. Interpret expressions that represent a quantity in terms of its context. • Interpret parts of an expression, such as terms, factors, and coefficients. • Interpret complicated expressions by viewing one or more of their parts as a single entity.	I can interpret expressions within a context.	• Interpret parts of an expression, such as terms, factors, and coefficients in terms of a given context. • Give a context and interpret complicated expressions by viewing one or more of their parts as a single entity.
2. Create equations that describe numbers or relationships. Create equations and inequalities in one variable and use them to solve problems. 3. Solve linear equations and inequalities in one variable, including equations with coefficients represented by letters. 4. Understand solving equations as a process of reasoning and explain the reasoning. Explain each step in solving a simple equation as following from the equality of a number asserted at the previous step starting from the assumption that the original equation has a solution. 5. Rearrange formulas to highlight a quantity of interest, using the same reason as in solving equations. For example, rearrange Ohm's law $v = IR$ to highlight resistance R. 6. Use units as a way to understand problems and to guide the solution of multistep problems; choose and interpret units consistently in formulas; choose and interpret the scale and the origin in graphs and data displays. 7. Define appropriate quantities for the purpose of descriptive modeling.	I can create, solve, and reason with linear equations and use appropriate units in context. I can create, solve, and reason with linear inequalities and use appropriate units in context.	• Create one-variable equations and use them to solve problems. • Create one-variable inequalities and use them to solve problems. • Use units as a way to understand and solve multistep problems. • Choose and interpret units in formulas and graphs. • Define quantities. • Explain the steps used to solve simple equations. • Rearrange formulas to highlight a quantity of interest.

Figure 1.4: Sample essential learning standards for an algebra 1 linear equations unit.

How to Implement Essential Learning Standards Routines

As you and your team design daily lessons, you also need to consider how you are going to use the daily learning targets in each lesson to communicate the learning pathway to students. The daily learning targets must be clear, and the use of them communicates the daily progression of how students become proficient in the essential learning standard. Together, the daily learning targets connect to the full essential learning standards for the unit.

However, sometimes the lesson may be best served if you delay *when* to reveal the essential standard to the students. It is possible that due to the exploratory nature of the lesson, you might want to wait. Regardless of when you reveal the learning target, be sure to set the context for how the learning target fits into the student progression for learning (what we have learned previously, what we will be learning later on, and how the standard for this lesson fits *now*—today).

When learning targets are clear for students, regardless of when they are introduced in a lesson, students can explain why they are doing a mathematical task, how the task is related to the unit or other areas of mathematics, and what the *why* is behind the learning.

Learning targets detail the learning progression of the essential standards for the unit, and help your students answer connecting questions such as, What have we learned so far? What are we learning now? And what will we be learning in the future?

Sharing learning targets with students happens in several ways. You can post the learning target or essential questions at the opening of the lesson, and then refer students to the learning target several times during the lesson. You can list the learning targets on the independent practice and connect those targets in writing to the formative assessment and feedback cycle. Learning targets and essential questions are part of the lesson-closure routines explored in more detail in chapter 6 (page 85).

In the following personal story, Jennifer Deinhart explains how teachers and teams can use student-friendly learning targets to help students understand the *why* of the lessons.

Personal Story  **JENNIFER DEINHART**

For elementary teachers, communicating about mathematics learning involves teachers as facilitators using language that makes sense to children and incorporating tools or visuals that support the language of the learning targets. We use student-friendly goal cards created by the teacher team to show a progression of learning in pictorial ways (see figure 1.5). These images and *I can* statements transfer to posters or can be projected during learning for specific teaching purposes. Teachers are referring to these learning targets while modeling during focus lessons, during student discourse when strategies are being shared during the prior-knowledge routine, and as a critical part of small-group instruction when students are engaging in next steps or trying out more efficient procedures.

The goal card clarifies the learning targets in the moment by giving students examples of what the learning looks like. This tool engages students in real-time reflection as we celebrate learning that has been mastered and also determine next steps. Using both common formative assessment tasks and common goal cards together allows teacher teams to monitor progress, effectively plan for student re-engagement in core instruction, and even communicate with parents about essential mathematics learning.

I know addition doubles facts within 20.	I can skip count by 2s, 5s, and 10s.	I can use repeated addition to count groups.	I can recall multiplication facts with 2s, 5s, and 10s.	I use 2s, 5s, and 10s facts to solve unknown facts.	I can double numbers within 100.	I can recall facts through 12 × 12.	I know prime, composite, and square numbers.
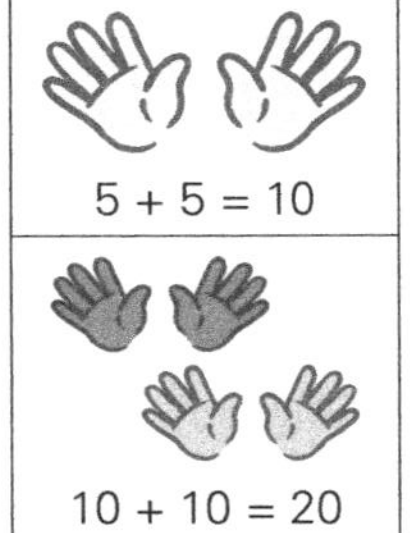 5 + 5 = 10 10 + 10 = 20	 "2, 4, 6, 8, 10, 12!"	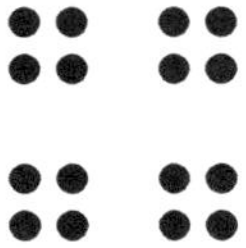 4 + 4 + 4 + 4	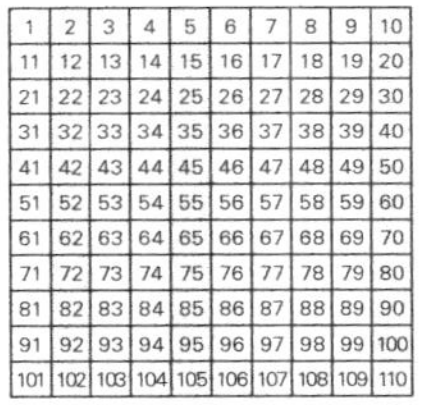	Use 5 × 5 = 25 to determine 6 × 5. Show your thinking.	Doubling 2 × 8 = 16 4 × 8 = 32 8 × 8 = 64	12 × 12 = 144 	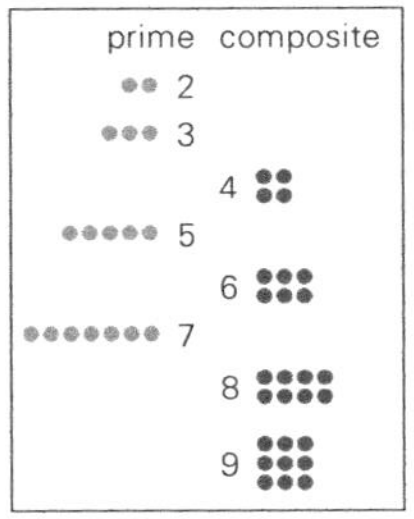 prime composite 2 3 4 5 6 7 8 9
I know subtraction doubles facts within 20.	**I can skip count back by 2s, 5s, and 10s.**	**I can use repeated subtraction to find fair shares.**	**I can recall division facts with 2s, 5s, and 10s.**	**I can use multiplication to solve unknown facts.**	**I can halve numbers within 100.**	**I can recall facts through 144 ÷ 12.**	**I can apply rules of divisibility.**
9 + 9 is 18 So I know that 18 – 9 is 9	What are the first two numbers? ?, ?, 35, 40, 45	I have 20 cookies. Each friend gets 4 cookies. How many friends will get cookies? 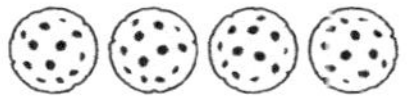	14 ÷ 2 = ? 20 ÷ 5 = ? 50 ÷ 10 = ?	54 ÷ 9 = ? 9 × 5 = 45 9 × 1 = 9 9 × 6 = 54	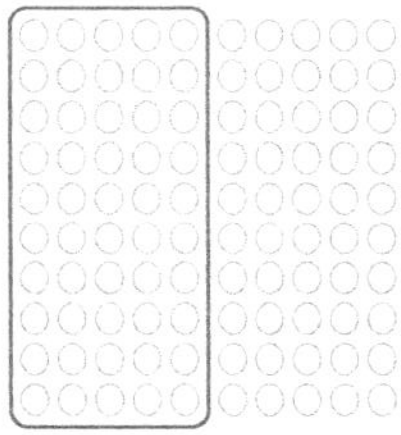	 Multiplication	A number is **divisible** by **6** if it is **divisible** by both **2** and **3**.

Figure 1.5: Sample elementary multiplication and division goal card.

For additional examples of student self-assessment forms, see *Mathematics Assessment and Intervention in a PLC at Work, Second Edition* (Schuhl, Kanold, Toncheff, et al., 2024).

The daily learning targets also drive the student-led closure that is explored more in chapter 6 (page 85).

When you and your team unpack your unit's essential learning standards into daily learning targets, you support your unit-planning work as you build your unit calendar and decide the number of days needed for each lesson and the full unit of instruction in order to fully develop student learning for each essential learning standard in the unit. For more information on how your team creates unit calendars, visit **go.SolutionTree.com/MathematicsatWork** or scan the QR code in this book to see protocols and examples from the *Mathematics Unit Planning in a PLC at Work* series (Schuhl, Kanold, Barnes, et al., 2021; Schuhl, Kanold, Deinhart, et al., 2021; Schuhl, Kanold, Deinhart, Larson, & Toncheff, 2020; Schuhl, Kanold, Kanold-McIntyre, et al., 2021).

Consider your current practices and complete the teacher reflection.

TEACHER *Reflection*

What are some of the ways you and your team have created a context for the essential learning standard for a lesson? How has this helped your students understand why the learning target or essential question is necessary for the lesson?

Reflection on Practice

When you work with your team to examine the essential learning standards for the unit and then determine the aligned learning targets for your daily lessons, you begin to ensure equitable learning expectations and experiences for the students in classrooms across your team.

Take a few moments to reflect on how your team develops understanding of each essential learning standard for the mathematics unit and the impact of those standards on your daily learning targets. Consider how many days of instruction you will need for each essential standard based on your understanding of the breakdown for the learning targets.

Remember, answering PLC critical question 1—What is it we want each student to know and be able to do?—is where each lesson starts and ends. Knowing your *why* for the lesson helps clarify the tasks and activities you choose and how you choose to engage students in the learning process. Knowing your *why* addresses the first critical question of a PLC (DuFour et al., 2016): What do we want all students to know and be able to do?

TEAM RECOMMENDATION

Understand the Essential Learning Standards—The *Why* of the Lesson

- As a grade-level or course-based team, identify the common essential learning standards for the unit and prepare the corresponding essential learning targets for each lesson.
- Use the essential learning standards for the unit to help clarify the overarching unit topics, write them as *I can* statements to aid student understanding, and discuss what students must know and be able to do (the daily learning targets) in each lesson to be considered proficient for the standard.

Knowing the *why* of a mathematics lesson comes from knowing the context of the lesson, and connects students to the relevance of the standard as important mathematics to learn. When you and your students reflect on which standards led into the current mathematics lesson, and how the current lesson can help their learning of mathematics a few weeks from now, you are creating a connecting storyline for them: *How do the standards taught in this lesson fit into the mathematical story arc from one year to the next, one unit to the next, or one lesson to the next?* In chapter 2 (page 23), you will discover how prior-knowledge routines prepare your students for entering the new knowledge of the mathematics lesson for the day, allow you to gather diagnostic information about their level of readiness to learn, and help create a meaningful context for the standard they are learning during lessons that lie ahead.

Visit **go.SolutionTree.com/MathematicsatWork** for free reproducible versions of tools and protocols that appear in this book, as well as additional online only materials.

CHAPTER 2

Prior-Knowledge Routines

> Study findings suggest that learning of new content is supported by activating a conceptually relevant prior-knowledge sequence that helps connect the prior knowledge with the new knowledge.
>
> —*Pooja G. Sidney and Martha W. Alibali*

Your mathematics lesson should always begin with a connection to prior knowledge. Your connection to each student's prior knowledge as they enter into the lesson for the day establishes the foundation for the grade or course essential learning standard. Prior-knowledge routines can include warm-up tasks or activities, examples, number talks, or a set of problems that support making connections between previous and new learning.

Why should you ensure every mathematics lesson begins with a prior-knowledge routine? The simple answer: *student perseverance.* When you monitor students' initial response to entry concepts into the lesson, they are more likely to persevere longer as the lesson begins (Hattie & Yates, 2014).

Students need to know their current understanding is honored and valued as they make connections from what they already know and what they are about to learn. Also, by not focusing on student deficits, you begin a lesson by identifying your students' prior-knowledge strengths and build your next instructional steps based on their individual points of power (Kobett & Karp, 2020). Moreover, since the prior-knowledge routine is generally a review, it allows you to gather evidence of overall student readiness to begin the lesson.

Additionally, a prior-knowledge activity allows you to establish context (or purpose) to the lesson, and provides an excellent opportunity to draw on multiple resources of knowledge in order to make instruction more meaningful to students (Aguirre et al., 2013).

You can use prior-knowledge routines to create a much-needed context for the *why* of the lesson. You can explain to students about standards they learned in the past and make connections from today's lesson to that prior knowledge. Your "creating context" discussion, which can come from the prior-knowledge problem, prepares students for the lesson you are about to begin. Though it might be more efficient to *tell* your students what they have previously learned and how it connects to the day's lesson, *asking* your students to engage in revisiting prior knowledge through a daily routine provides a context for your students, and for you, to see and hear their understanding of the connections. Prior-knowledge activities and tasks support how you envision the lesson's learning target fitting into the mathematics learning progression (or progressions) of the unit, and they reflect your willingness to use mathematical outcomes to focus students' learning (NCTM, 2014).

During 2020 and 2021, the COVID-19 pandemic disrupted student learning in mathematics—a vertically connected curriculum. Working with your teacher team to combine the use of prior-knowledge routines (what the students recall and connect to) with the mathematics learning progressions will provide just-in-time support as students close unintentional gaps in learning the new standard for the daily lesson. Additionally, the National Council of Teachers of Mathematics and National Council of Supervisors of Mathematics (2020) state:

> Educators should view students in terms of their strengths, not weaknesses, and avoid the urge to immediately reteach all the skills we think students should have learned before arriving at school this fall. It is more productive for teachers to think of learning opportunities that are most important for students in relation to the mathematics learning progressions. (p. 7)

As teams develop a stronger sense of the learning progressions, teachers are able to strategically design prior-knowledge routines that better support student understanding.

Design of Prior-Knowledge Routines

You need to be mindful of the amount of lesson time that will be needed for your prior-knowledge routines. In most mathematics lessons, warm-up activities that activate prior knowledge are student-centered activities generated by either a mathematical task, a problem set, or a discussion prompt you provide to your students as the lesson begins.

A prior-knowledge routine should not use a lot of lesson time; no more than five to ten minutes is appropriate. If the activity is more of an exploration-based prior-knowledge task (a mathematical task involving investigation rather than questions to answer), then slightly more time might be necessary. In the following personal story, Sarah Schuhl references a conversation with a team about the amount of time spent on a prior-knowledge task while also discussing the nature or intended purpose of the task.

Personal Story  SARAH SCHUHL

I worked with a third-grade team to collaboratively plan a mathematics lesson related to areas of shapes composed of rectangles. When I asked how the lesson would begin with students, the team indicated it always started every lesson with a worksheet from a program that asked students ten random mathematics questions. We examined the questions. The questions for the area lesson ranged from elapsed time to addition with an algorithm to fraction equivalence to pattern recognition. I asked how much of the one-hour lesson was devoted to this warm-up each day. The answers ranged from fifteen to twenty-five minutes. Several teachers explained that students did not know the concepts, and so they retaught the ten problems before starting the lesson. They further explained that the area lesson would be difficult and they wished they had more time.

After much discussion, the teachers finally agreed to write their own warm-up activity. They wrote one question asking students to find the perimeter and area of a rectangle with only two sides labeled so students would practice area of a rectangle as well as be reminded that they can find missing side lengths when they are not included in the diagram. The teachers wrote a second question giving students the area of a rectangle with its base and asking students for the height, again reinforcing the idea that missing side lengths can be determined. They agreed to give students five minutes to work on the problems in their groups and then have two groups share their solutions before launching the lesson with a shape composed of two rectangles and asking students to find its area in their groups.

After the lesson, the teachers were excited by the evidence of student learning in the lesson. They felt the prior-knowledge routine was quick and activated the knowledge needed to maximize the time spent practicing areas of complex shapes. While spiral review is important, they decided they would create the warm-ups to prepare students for the learning of the day and might occasionally include one spiral review question, if needed. The team decided it would also work to include the spiral review questions in their homework assignments.

When you are selecting prior-knowledge tasks to support essential learning standards for a lesson, it can be helpful to reference the prior standards as well. These could be standards from a prior grade (or course) to see the level of depth explored the year before. Alternatively, these could be standards from a prior unit. Either way, the goal is to help students connect to the content of the lesson in order to help them make meaning. This helps to clarify where your instruction needs to start and the prior learning you want to connect with through the warm-up activity.

Figures 2.1, 2.2, 2.3 (page 26), and 2.4 (page 27) provide examples of prior-knowledge routines for grades 2, 5, and 7 mathematics and a high school algebra 1 course. They are based on a standard from the prior grade level. However, keep in mind that sometimes the prior-knowledge mathematical task is from a previous unit or lesson and may not be from a prior grade level or course. As you design your prior-knowledge routine, make sure the task or activity reflects knowledge most connected to the essential learning standard for the mathematics unit, and the learning target you are about to ask students to engage in learning for the day.

Take some time to closely examine the prior-knowledge task most appropriate for your grade level.

<table>
<tr><th>Grade or Course</th><th>Grade-Level or Course-Based Standard</th><th>Prior-Knowledge Standard From Prior Grade, Course, or Unit</th></tr>
<tr><td>Grade 2</td><td>Formal unit standard:
Use addition and subtraction within 100 to solve one- and two-step word problems involving situations of adding to, taking from, putting together, taking apart, and comparing, with unknowns in all positions.
Essential learning standard:
I can use addition and subtraction within 100 to solve one- and two-step word problems.
Daily learning target:
Students will be able to use addition and subtraction to solve one-step word problems involving adding to and putting together.</td><td>Grade 1:
Use addition and subtraction within 20 to solve word problems involving situations of adding to, taking from, putting together, taking apart, and comparing, with unknowns in all positions.</td></tr>
<tr><td colspan="3">Prior-knowledge task sequence:
1. Pat has eight red flowers and two yellow flowers in a vase. How many flowers are in the vase altogether? Show how you know.
2. There are twelve flowers in a vase. Five are white, and some are purple. How many are purple? Show how you know.</td></tr>
<tr><td colspan="3">Explanation of task:
By starting with tasks that include numbers under 20, teachers can learn whether or not students understand one- and two-step problems and the strategies that go along with solving them. These prior-knowledge examples specifically focus on putting together and taking apart. Teachers have the opportunity to see how students show their work, especially for question number two. Do students use addition, subtraction, or do they add on?</td></tr>
</table>

Figure 2.1: Grade 2 prior-knowledge routine example.

<table>
<tr><th>Grade or Course</th><th>Grade-Level or Course-Based Standard</th><th>Prior-Knowledge Standard From Prior Grade, Course, or Unit</th></tr>
<tr><td>Grade 5</td><td>Formal unit standard:
Find whole-number quotients of whole numbers with up to four-digit dividends and two-digit divisors, using strategies based on place value, the properties of operations, and/or the relationship between multiplication and division. Illustrate and explain the calculation by using equations, rectangular arrays, and/or area models.
Essential learning standard:
I can find whole-number quotients of whole numbers using multiple strategies.
Daily learning target:
Students will be able to find whole-number quotients of three-digit dividends and two-digit divisors using multiple strategies and explain the calculation using an area model or rectangular array.</td><td>Grade 4:
Find whole-number quotients and remainders with up to four-digit dividends and one-digit divisors, using strategies based on place value, the properties of operations, and/or the relationship between multiplication and division. Illustrate and explain the calculation by using equations, rectangular arrays, and/or area models.</td></tr>
</table>

Figure 2.2: Grade 5 prior-knowledge routine example.

continued →

Prior-knowledge task sequence: 1. Gabriella makes 7 waffles for breakfast. She has 42 berries to put on top of her waffles. She will put an equal number of berries on each waffle. How many berries will Gabriella put on each waffle? Show your work and write the equation. 2. Now suppose Gabriella has 52 berries. What will happen? How many will she put on each waffle and why? 3. Solve 3,476 ÷ 7. Show your work.
Explanation of task: For the first day of grade 5 instruction on division, knowing the grade 4 standard is four-digit dividend and one-digit divisor, it would be beneficial to start with a task that the majority of grade 5 students will be able to connect to without getting too caught up in errors. This will still give the teacher good insight into student misconceptions and their overall understanding of division. Before asking a rote division question, it helps for you to know if your students can pick up division in context of a word problem.

Grade or Course	Grade-Level or Course-Based Standard	Prior-Knowledge Standard From Prior Grade, Course, or Unit
Grade 7	**Formal unit standard:** Compute unit rates associated with ratios of fractions, including ratios of lengths, areas, and other quantities measured in like or different units. **Essential learning standard:** I can compute unit rates associated with ratios of fractions. **Daily learning target:** Students will be able to compute unit rates associated with ratios of fractions.	**Grade 6:** Understand the concept of a unit rate $\frac{a}{b}$ associated with a ratio $a:b$ with $b \neq 0$, and use rate language in the context of a ratio relationship.
Prior-knowledge task sequence: Two pizza restaurants are running specials on large pizza orders. Tony's Pizza is selling 3 pizzas for $36.00. Pizza Place is selling 5 pizzas for $55.00. Which pizza restaurant has the best deal? Explain your thinking.		
Explanation of task: Before you dive into the grade 7 standard about computing unit rates, it could be helpful to know what students remember about unit rates from grade 6. The question being asked is challenging because students have to recall the information and create their own example, which requires them to work backward and create. As students are working, the conversations they will be having with their peers should reveal their understanding of the topic. If students struggle through this warm-up, even with a numerical example, then the teacher knows students need to start at the grade 6 level before they can jump into the expectation of the grade 7 standard. If students demonstrate an understanding of unit rates, then the teacher knows to start with more challenging tasks for the lesson.		

Figure 2.3: Grade 7 prior-knowledge routine example.

In the grade-level examples provided in figures 2.1, 2.2, 2.3, and 2.4, you will notice the use of a sequence of questions to gather evidence of prior knowledge. However, it is also possible that the prior-knowledge routine is an exploration-type question that students work on together and investigate, based on prior knowledge, without having to answer a series of questions. These types of prior-knowledge routines can help students establish a *why* for the day's learning targets as well.

For example, in a grade 3 lesson introducing division, you might start with some multiplication problems that ask students to think about the number of groups and the number of objects within the groups. You could ask students to create a word problem context for the product of 4 × 6 or 9 × 7. They could draw or write out verbal examples. Then as you introduce the concept of division, you can tie in the same language used in the prior-knowledge routine to help students make the connection between division and multiplication.

Grade or Course	Grade-Level or Course-Based Standard	Prior-Knowledge Standard From Prior Grade, Course, or Unit
Algebra 1	**Formal unit standard:** • Understand that the graph of an equation in two variables is the set of all its solutions plotted in the coordinate plane, often forming a curve (which could be a line). • Graph linear functions expressed symbolically and show key features of the graph, by hand in simple cases and using technology for more complicated cases. **Essential learning standard:** I can graph linear and nonlinear functions. **Daily learning target:** • Students will be able to graph different functions by hand. • Students will be able to distinguish between linear and nonlinear functions.	**Grade 8:** • Interpret the equation $y = mx + b$ as defining a linear function, whose graph is a straight line; give examples of functions that are not linear. • Derive the equation $y = mx$ for a line through the origin and the equation $y = mx + b$ for a line intercepting the vertical axis at b.

Prior-knowledge task sequence:

1. Write the equation for the line graphed below.

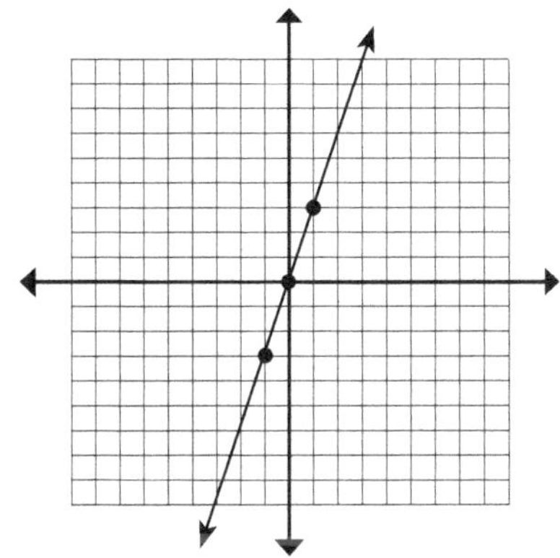

2. Write the equation for the line graphed below.

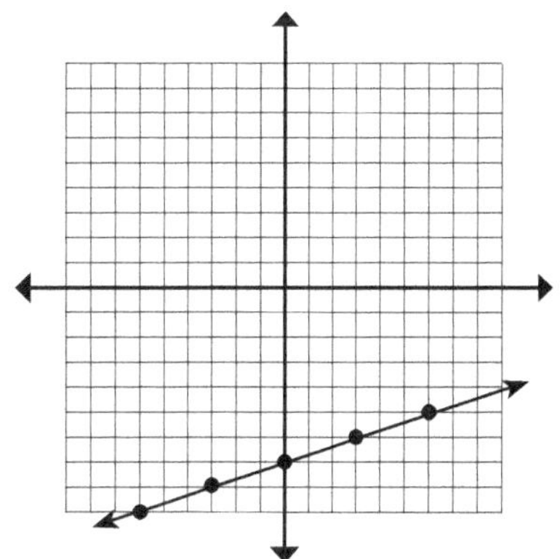

3. What is different about these two graphs? What do you notice about the intercepts and the slopes?

4. Graph an example of a nonlinear function. Be prepared to share with a neighbor.

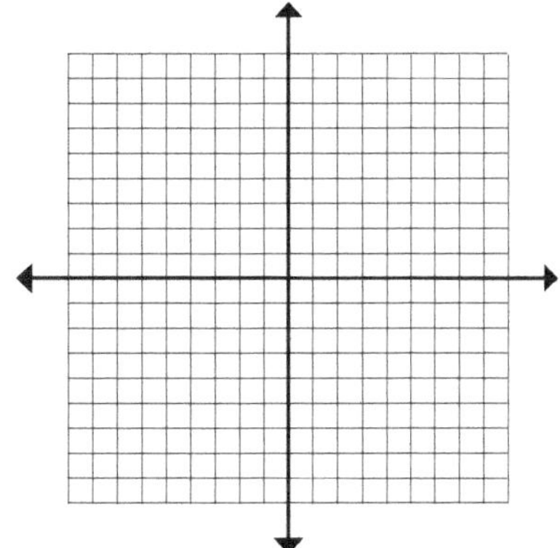

Explanation of task:

The algebra 1 high school standards for this lesson relate to graphing. Given the knowledge students will have based on grade 8 standards, this warm-up starts with the basics of graphing and the vocabulary tied to graphing in order to determine what students remember and can apply. Sample vocabulary might consist of constant rate of change (slope), intercepts, domain, and range. The vocabulary word *function* is new to the high school standards.

Figure 2.4: High school algebra 1 prior-knowledge routine example.

In an algebra 1 course, as a prior-knowledge task, you can ask students to solve a system of equations written in standard form, knowing the only strategy they currently employ for solving a system is graphing. Graphing presents a long and tedious process to the students. Thus, the activity builds an *efficiency reason* to learn and identify a more effective strategy like substitution or elimination. It starts to build a *representation* context for why you would choose between various solution pathways with or without technology, such as graphing, substitution, or elimination methods.

Your team discussion and identification of prior-knowledge standards for the current unit provide information needed to create effective prior-knowledge tasks at the start of each lesson. For more information, see the *Mathematics Unit Planning in a PLC at Work* series.

Now, take a few moments to consider the teacher reflection for the warm-up activities you choose for your lessons and how you determine which tasks and problems to use.

TEACHER *Reflection* 

What is the best prior-knowledge routine you use in your lessons? Do you prefer to use mathematical tasks, discussion prompts, or both for your warm-ups?

As a team, identify an essential learning standard in your next unit of study. You can use figure 2.5, the prior-knowledge routine-planning tool, to identify the prior grade or unit essential standard you might need to reference, and choose either a mathematical task or a discussion prompt that ties into the prior-knowledge skills that will help students access the essential learning standard for the lesson. In the explanation section, make some notes about why you chose the specific prior-knowledge task or tasks.

At first, it is helpful to write an explanation, but as you progress, you may find it is just as beneficial to verbally discuss the reasoning.

As you and your team consider the prior-knowledge routines, it's important to remember that the mathematical tasks you choose and the discussion prompts you provide need to connect to the conceptual understanding you are trying to build. This is why it is so important to reference prior standards that align to the same strand in the same grade or from a previous grade. The connections help you and your students understand and observe the vertical connections and progressions between the various units of content across grades and courses.

TEAM RECOMMENDATION

Plan Prior-Knowledge Routines

- When planning prior-knowledge routines, be sure to choose mathematical tasks and discussion prompts that connect to the conceptual understanding you are trying to develop for the essential learning standard and learning target that day.
- Design prior-knowledge routines to help connect to the why of the lesson, paint a picture of where students are headed in the lesson, and develop student perseverance during the lesson (by reminding them throughout the lesson how chosen activities connect to the learning standard).

Up next, you will explore how to implement the prior-knowledge routines.

<table>
<tr><th>Grade or Course</th><th>Grade-Level or Course-Based Standard</th><th>Prior-Knowledge Standard From Prior Grade, Course, or Unit</th></tr>
<tr><td></td><td>Formal unit standard:

Essential learning standard:

Daily learning target:</td><td></td></tr>
<tr><td colspan="3">Prior-knowledge task sequence:</td></tr>
<tr><td colspan="3">Explanation of task:</td></tr>
</table>

Figure 2.5: Prior-knowledge routine-planning tool.

Visit ***go.SolutionTree.com/MathematicsatWork*** *for a free reproducible version of this figure.*

How to Implement Prior-Knowledge Routines

After your team designs the prior-knowledge routine, you then need to plan how to implement the routine. To begin, as a team, use figure 2.6 to discuss how you currently implement prior-knowledge routines that support student engagement while also helping students connect to the learning target for the day's lesson.

Guidelines to Consider

As your collaborative team reflects on your responses in figure 2.6, there are four helpful guidelines to consider when implementing the prior-knowledge routines.

1. **Always use a small-group focus:** Small groups provide students with an opportunity to work together and practice communicating their ideas to review a concept from the prior lesson, unit, or year. Small groups also create a natural opportunity for students to provide each other with feedback and re-engage by reviewing content before exploring new content for the lesson. You should tour the students' peer-to-peer discussions to see and hear student understanding and provide small-group discourse feedback as students need it.
2. **Provide higher-level-cognitive-demand tasks and prompts:** The mathematical tasks or discussion prompts you choose for the prior-knowledge routine should generally be rigorous and promote mathematical thinking and student understanding on previously learned standards. Prior-knowledge tasks should require your students to reason, justify, or problem solve—tasks that go beyond demonstrating a routine skill.

   The purpose of the prior-knowledge routine is to promote connections to the essential learning standard or new knowledge to be taught and the critical thinking to come ahead in the lesson. Try to avoid simple rote memorization of routine tasks; rather, present a question or task that seeks to determine what students *understand*.
3. **Structure a clear routine for the start of class:** There should be evidence the prior-knowledge task is built around a carefully selected mathematical task or prompt with a well-organized and understood routine for how your students are to proceed, engage, and interact with each other. The prior-knowledge task should be readily available to students as class begins, with clear directions and prompts for how to proceed and share their thinking with one another. You should use no more than five to ten minutes of the overall lesson time as students respond to the discussion prompt or the mathematics problems you provided for connecting their prior knowledge and understanding.
4. **Do not "go over" the prior-knowledge tasks in class:** As you walk around the room and observe student teams successfully engaging in the prior-knowledge routine, allow them to discuss the mathematics task or prompt with their peers and with feedback from you as you determine their readiness for the lesson. Do not spend time "going over" the task in class. Students can review answers you supply to check their work themselves. Or, as you walk around the room monitoring students, you may want to reveal a few student solutions to share on the document camera or other public display for students to review while they are discussing the prompt from the activity.

If you observe students struggling during the prior-knowledge routine, this is a great opportunity to reassess your next steps before you start the lesson. If there are just a few students struggling, then you may strategically pull a small group of students aside during an appropriate time during the lesson. If you see the majority of the class struggling, then you may need to redirect or shift the lesson. You will need to ask questions, provide insight, or give students an additional scaffolding prompt to help them re-engage in the mathematical task.

Remember that your students come to the mathematics lesson with a broad range of pre-existing knowledge and skills. How well they persevere through, process, and integrate the new information from your

Directions: As a teacher team, discuss your current process for implementing a prior-knowledge routine using the questions that follow.

1. What are your current prior-knowledge routines? And how do the routines create a context for the learning target of the day? Share your ideas as a team.

2. How do you structure the student discourse with peers during the routine (since it is a review activity)?

3. In what ways do you use the prior-knowledge routine to assess student readiness for the lesson?

4. How do students receive feedback during the prior-knowledge routine (from students, from the teacher, or from both)?

Figure 2.6: Teacher team discussion tool—Prior-knowledge routine process.

Visit ***go.SolutionTree.com/MathematicsatWork*** *for a free reproducible version of this figure.*

daily lesson is influenced by the connections you help them make to previous learning during the prior-knowledge routine.

Since this lesson-design element is designed to assess a prior-knowledge mathematics concept or skill, a natural outcome of the activity is to reveal the new essential standard and learning target for the lesson that day once the routine is completed. It creates the *why* for the lesson and the learning progression context for the students.

For example, you could ask your students a question that sets up the context for the progression of the next learning standard connected to the day's lesson: "What do you remember about base ten from second grade? Well, today in this third-grade lesson on base-ten numbers, we are going to study . . ." or "Recall our discussion about equivalent expressions from the previous unit. What is meant by the word *equivalence*? In this lesson today, we are going to study . . ." or "Do you remember that linear functions represent a constant rate of change? Well, in today's lesson, we are going to study . . ."

TEAM RECOMMENDATION

Implement Prior-Knowledge Routines

- As a team, understand how prior-knowledge routines and mathematical tasks provide the context for why the essential standard for the day's lesson is relevant and why students need to learn it.
- Prior-knowledge routines create an entry point into your lesson by activating students' knowledge of connecting concepts.
- The prior-knowledge routine is a great way to support student engagement in a collaborative process and develop student confidence.
- The prior-knowledge task provides you an opportunity to assess student readiness for the lesson.

Reflection on Practice

In general, moving out of the prior-knowledge routine and into a public declaration of the purpose and context for this day's lesson—to learn the essential learning standard of the unit and the learning target or essential question for the day—is the best process to follow. Remember that to answer PLC critical questions 3 and 4—How will we respond in class when some students do not learn? and How will we extend the learning in class for students who are already proficient?—begins with PLC critical question 1, What do we want all students to know and be able to do *today*?

So what is next? Once you and your team discuss how you implement prior-knowledge routines (it can be very robust and varied), be sure to discuss the strategies you use to support mathematical language during your daily lessons. You and your team also need to focus on the mathematical language, both words and notations, incorporated into your lesson and unit planning, and on the potential meaning-making barriers your students will face when you (and they) fail to address vocabulary as part of the lesson.

The next chapter explores your intentional design for the use of mathematical language and notation in great detail.

Visit **go.SolutionTree.com/MathematicsatWork** for free reproducible versions of tools and protocols that appear in this book, as well as additional online only materials.

CHAPTER 3

Mathematical Language Routines

Students need to know the meaning of mathematics vocabulary words—whether written or spoken—in order to understand and communicate mathematical ideas.

—*Rheta N. Rubenstein and Denisse R. Thompson*

Mathematical language is the language students use to make sense of and communicate the mathematics they are learning. Students will listen, read, write, and speak while engaging in solving mathematical tasks. Mathematical language includes knowing the meaning of and using words, symbols, notations, and abbreviations, which are important when students are learning essential standards.

Mathematical language is important to students' ability to communicate verbally and in writing as they evaluate progress on mathematical tasks experienced during a lesson, analyze their own thinking, construct an argument, or justify their reasoning with confidence to you and their peers. Despite this importance of mathematics vocabulary and notation routines, mathematical language is an often-overlooked aspect of the lesson-design process and implementation.

This is understandable to some extent, as the typical mathematics lesson tends to focus on the mathematics the student needs to do in order to learn the standard. And yet words like *denominator, contradict, area, mean ratio, equivalence, tangent,* and *slope field* can confuse many students when it comes to the intent of the lesson.

In order for students to communicate about their ideas orally and in writing, they need a strong mathematical language. According to Meir Ben-Hur (2006), senior international lecturer for the Feuerstein Institute, "Students who lack the formal language of mathematics have difficulties reasoning and communicating about mathematics" (p. 67). Multilingual learners also need to engage in mathematical discourse. Students learning English need opportunities to learn mathematics vocabulary, engage in discourse, make conjectures, and explain their thinking (Civil & Turner, 2014; Zahner, Pelaez, & Calleros, 2020; Zwiers et al., 2017).

As a collaborative team, you and your colleagues discuss the vocabulary and notations students need to learn when you are planning your units together. Your team agrees on the mathematical language to use to create student agency for learning. These agreements also establish consistency across all lessons for the unit as your students access tasks and clearly communicate their mathematical reasoning for the unit, whether or not they are enrolled in same class. For more information, see the *Mathematics Unit Planning in a PLC at Work* series.

At your next team meeting, ask each team member to share how they currently support the daily development of mathematical language—vocabulary and notations—*during* a lesson. You can use the teacher reflection as a prompt for the discussion.

TEACHER *Reflection* 

What strategies and tools do you currently utilize on a consistent basis to ensure each student understands the lesson's mathematical language (both vocabulary and notations)?

To help your students learn the content tied to the essential learning standards for the unit, you intentionally plan time for mathematical language instruction, on both words and notations, to support the overarching goal of your instructional plan. Mathematics is communicated by means of a powerful language whose vocabulary must be learned. The ability to reason about and justify mathematical statements is fundamental, as is the ability to use terms and notation with appropriate degrees of precision (Ball et al., 2005).

Effective Mathematical Language Routines

Joan M. Kenney, Euthecia Hancewicz, Loretta Heuer, Diana Metsisto, and Cynthia L. Tuttle (2005) indicate that mathematics textbooks expect students to use a vocabulary for understanding both words and symbols:

> Research has shown that mathematics texts contain more concepts per sentence and paragraph than any other type of text. They are written in a very compact style; each sentence contains a lot of information, with little redundancy. The text can contain words as well as numeric and non-numeric symbols to decode. There may also be graphics that must be understood for the text to make sense. (p. 11)

Students need to learn not only the words but also the symbols that go along with the mathematics. Providing definitions during instruction is not sufficient for the long-term demonstration of student understanding and correct use of mathematical language.

Mary Lee Barton and Clare Heidema (2000) further indicate:

> Reading mathematics means decoding and comprehending not only words but also mathematical signs and symbols. . . . Students must switch between the skills they use to decode words and the skills needed to decode signs and symbols. . . . Students need to learn the meaning of each symbol and to connect each symbol, the idea the symbol represents, and the written or spoken words. (p. 11)

The mathematical vocabulary students encounter often causes confusion and limits their access to understanding of the mathematics standard for the unit. For example, consider the word *meter*. In a lesson or unit on measurement, you start students with the root word and then continue on to other words such as *millimeter*, *centimeter*, *decimeter*, and *kilometer*. However, later on, during a geometry unit, you teach the word *perimeter*. Students might ask, "How many in a peri?" This is an example of where the vocabulary and language of mathematics can get confusing for some students, especially if your lesson design is not intentional about supporting student meaning making for their own organization of ideas. Intentionally teaching students the origin of the prefix *peri-* can support mathematical meaning later in the lesson.

One way to help organize different types of vocabulary words is to think about the words students encounter as a tiered system. In this three-tier system from Isabel L. Beck, Margaret G. McKeown, and Linda Kucan (2013), vocabulary words are classified into three tiers of organization.

- **Tier one** words are everyday, common words such as *more*, *less*, *high*, and *low*.
- **Tier two** words include vocabulary words that span multiple academic settings such as *analyze*, *evaluate*, *explain*, and *justify*.
- **Tier three** vocabulary words are content-specific words like *quotient*, *quadratic*, *polynomial*, and *denominator*.

Literacy expert Kimberly Tyson (2013) suggests "understanding the three tiers can help separate the 'should-know words (tier three)' from the 'must-knows (tier two)' and the 'already-known words (tier one).'" Based on these tiers, the majority of your vocabulary instructional energy should focus on tier three with a check-in on tier two words, as needed.

In figure 3.1, Rheta N. Rubenstein and Denisse R. Thompson (2002) provide examples of areas of challenge that may make vocabulary development difficult for your mathematics students. After you review the chart and examples as a team, think about an upcoming unit of mathematics instruction. What are the challenges with the vocabulary your students will encounter throughout the unit? What words might students struggle with the most? (Hint: They might be words that explicitly appear in the standards and directly relate to the mathematics content you will teach [tier three vocabulary], but they might also be more global, cross-curricular words like *evaluate*, *analyze*, and *explain* [tier two vocabulary].) As you read the examples in figure 3.1, use the questions provided in the teacher reflection (page 36).

Area of Challenge	Possible Examples	Team Vocabulary Reflection for the Current Mathematics Unit
Mathematics and everyday English share some words, but they have different meanings in the two contexts or the mathematics meaning is more precise.	• *Right* angle versus *right* answer • *Foot* as twelve inches versus *foot* as body part • *Difference* as the answer to a subtraction problem versus *difference* as a general comparison	
Some mathematics words are found only in mathematical contexts.	• Quotient • Denominator • Integer • Isosceles • Histogram	
Some words have more than one mathematical meaning.	• *Square* as a shape versus *square* as a number times itself • *Round* as a shape versus *round* as a number operation	
Some mathematical words are related, but students may confuse their distinct meanings.	• *Hundreds* and *hundredths* • *Factor* and *multiple* • *At most* and *at least* • *Solve* and *simplify*	
English spelling and usage have many irregularities.	• *Four* has a u, but *forty* does not. • Fraction denominators, such as *sixth*, *fifth*, *fourth*, and *third*, are written like ordinal numbers, but rather than *second*, the next fraction is *half*.	
Some mathematics concepts are verbalized in more than one way.	• *Skip count* versus *find the multiples* • *One quarter* versus *one-fourth* • *Solutions*, x-*intercepts*, and *roots*	
Some mathematical words are homonyms with everyday English words.	• *Sum* versus *some* • *Arc* versus *ark* • *Pi* versus *pie* • *Graphed* versus *graft* • *Whole* versus *hole*	

Source: Adapted from Rubenstein & Thompson, 2002.

Figure 3.1: Teacher team discussion tool—Categories of vocabulary challenges for students.
*Visit **go.SolutionTree.com/MathematicsatWork** for a free reproducible version of this figure.*

How do you help students demonstrate clarity regarding essential vocabulary words during the unit? There are many *informal* words, symbols, and phrases used every day during mathematics lessons. Although seemingly harmless, the lack of precision in communicating about the mathematics can be a sense-making detriment to students as the mathematics standards become more complex.

TEACHER *Reflection* 

What words do you use with students on a daily basis for your current unit of mathematics study? Which words or notations consistently cause students problems during the unit?

How precise is your use of mathematical language for each lesson, and how precise are your expectations for student precision?

As you design lessons that develop mathematical language, you will also need to consider what mathematical language routine will develop students' knowledge of the mathematical vocabulary and notation. A *mathematical language routine* is defined as:

> a structured but adaptable format for amplifying, assessing, and developing students' language. . . . These routines can be adapted and incorporated across lessons in each unit to fit the mathematical work wherever there are productive opportunities to support students in using and improving their English and disciplinary language. (Zwiers et al., 2017, p. 9)

It's just as important for you, the teacher, to use correct mathematical terms and notation as it is for students to use those terms correctly. For example, a high school teacher of mathematics who uses the word *corner* instead of *vertex* is not modeling the correct language that aligns to the standards. Neither is an elementary teacher who says the number is *getting bigger* instead of *increasing* in value. If you lack precision, you will make it more difficult for your students to be precise.

As students explain their reasoning and communicate with one another over more complex and conceptual mathematical tasks, their achievement does improve. "Knowledge of mathematics vocabulary affects achievement in mathematics, particularly in the area of problem solving," according to Robert Helwig, Marick A. Rozek-Tedesco, Gerald Tindal, Bill Heath, and Patricia A. Almond (1999, p. 113).

John Hattie, Douglas Fisher, and Nancy Frey (2017) indicate "what we say to students, as well as how we say it, contributes to their identity and sense of agency, as well as to their success" (p. 203).

Consider using the teacher reflection as a team activity as well, and discuss how you encourage classroom communication for vocabulary development.

TEACHER *Reflection* 

How do you currently include mathematical language as part of each daily mathematics lesson?

How do you intentionally plan for student communication that fosters accurate use of mathematics vocabulary and notations for all students, including your multilingual students?

When you establish strong mathematical language routines as part of your daily lesson and overall unit design, students will gain agency—their voice—in learning. When you address the mathematics vocabulary to equip all students with the confidence of using and understanding the language of mathematics as they discuss strategies to solve mathematics problems in class and embrace errors in their own thinking, then their learning of the essential learning standards will deepen. In her personal story, Jessica Kanold-McIntyre shares how to intentionally design space for teachers and teams to come to agreement on the common mathematical language during each unit.

Personal Story  JESSICA KANOLD-MCINTYRE

In most schools I work with, I hear a common theme when it comes to curriculum implementation. Teachers express concern that each teacher has their own language or words that they use with students; teachers need help building a more consistent mathematical language in their classrooms. Every time I hear this concern, I am reminded that not only is it important for a classroom teacher to build vocabulary instruction into their lesson-design process, but it is equally important to discuss key vocabulary for the unit as a grade-level or course-based team so there are common expectations across classrooms in the school (K–5) or a mathematics department (middle school and high school). In order to ensure equity for all students in a grade level or a mathematics course, students need to experience accurate and common mathematics vocabulary, including notation.

In my district's curriculum-writing process, we intentionally built in a place to record common academic vocabulary that we would all agree to teach for every mathematics unit. We created a systematic process in K–8 to ensure we were all in agreement on the common mathematical language for each unit. This component of the unit-planning process also transferred to our middle school courses we had in common with the high school to ensure we had consistent expectations with the high school faculty as well.

TEAM RECOMMENDATION

Include Mathematical Language Routines in Lesson Design

- In order for students to be able to justify, communicate, and reason mathematically, it is crucial they have a strong vocabulary of words, symbols, and notations.
- Multilingual students need opportunities to learn mathematics vocabulary, engage in discourse, make conjectures, and explain their thinking (Civil & Turner, 2014).

The first part of this chapter defines and highlights the importance of mathematical language routines when designing lessons. In the next part of the chapter, you and your colleagues will have a chance to explore *how* to implement mathematical language routines each day.

How to Implement Effective Mathematical Language Routines

In addition to the mathematical experiences every day, mathematics teams can utilize a variety of mathematical language routines to develop academic language and empower learners to maintain perseverance and engagement in daily lessons.

To begin, take time to reflect on your current mathematical language routines utilized within your mathematics instruction.

TEACHER *Reflection* 

How do you and your collaborative team members currently use mathematical language routines as a way to maintain student perseverance and engagement throughout a mathematics lesson and during the unit of instruction?

In the following personal story, Jessica Kanold-McIntyre shares her experience and growth with utilizing mathematical language routines.

Personal Story  **JESSICA KANOLD-MCINTYRE**

As a middle school mathematics teacher, I remember learning that vocabulary instruction was important. I would start a word wall with great intention, and then by the second or third unit of the year, I would forget about updating the words.

I tried to embed mathematical vocabulary instruction into my day-to-day lessons, but I did not realize the importance of *frequency*. I would embed some vocabulary into my lesson, provide an example or a definition, and expect students to understand and reference the word on their own.

As I have learned more about working with the mathematical language and reading needs of my students, I have discovered they need time to make their own meaning with new vocabulary while also being encouraged to use the words throughout the instructional process. This means that just having students use the glossary in the book to write the definitions of words or complete a crossword puzzle with vocabulary words is not enough.

In order to support equitable student experiences with mathematical language between teachers, it is important for you and your collaborative team to create common expectations around the mathematical language and identify the new vocabulary on a unit-by-unit basis. As part of the planning process at the beginning of a unit, you and your team should identify and discuss the vocabulary terms and notations to ensure consistent use of both for all students in the grade level or course.

To identify vocabulary terms for a unit, it is best to go back to the actual formal essential unit standards for direction. You should have written the vocabulary for the unit on your unit plan document. Then, as you create each lesson, note which vocabulary words you plan to highlight within the lesson, especially relating the vocabulary words to the possible challenges students may have with those particular words.

Take a moment to consider how you choose vocabulary for your instruction in the teacher reflection.

TEACHER *Reflection* 

How do you currently choose vocabulary for your units or lessons? Are the words you choose consistent with those of other teachers across your grade level or course?

When and How to Teach Vocabulary

You may be wondering, "When is the best time during a mathematics lesson to tackle vocabulary instruction?" It depends. In *Visible Learning for Mathematics*, Hattie and colleagues (2017) report that you can use vocabulary instruction at almost any time during a lesson. First, you can use it as a preteaching activity as the lesson begins and based on feedback you received during the prior-knowledge routine. Second, you can use it to reinforce just-in-time learning during the lesson as you need it. Third, you can reinforce vocabulary by formalizing the meanings of key words at the end of a lesson as part of the closure routine.

There are several activities or strategies you can use to inform your thinking about mathematical vocabulary instruction. Table 3.1 identifies various types of vocabulary challenges students may have and provides language strategies to address each issue as a way to support your instruction (Rubenstein, 2007).

Table 3.1: Vocabulary Challenges and Focused Strategies

Challenge	Focused Strategy
Some words are shared with everyday English, science, or other disciplines and may have distinct or more technical meanings within mathematics.	Be aware of potential confusion. Distinguish the technical from the everyday meanings. Help students understand what the terms share and the reason why the common language term was adopted for mathematics.
Some words are found only in mathematics.	Help students see the roots and origins of mathematics terms. Point out common English words with the same root. Help students see how the roots build the mathematical meaning.
Some words have more than one mathematical meaning.	Remind students of the multiple usages of words, and remind them to use context clues to know which is the intended meaning. Help students see why the word makes sense in each context.
Some words are learned in pairs that often confuse students.	If possible, separate the learning of the two terms so that students understand one word well before introduction of the second word. Continue to use word origins and relate each of the two terms to everyday English words. Acknowledge the challenge, and have students double-check one another when they use the words.
Some words sound like others (homonyms and near homonyms).	Say the words clearly. Spell them. Distinguish them. Use each in its particular context. Draw a picture.
Sometimes modifiers change meanings of words in critical ways.	Have students explore the unmodified term, the modifier, and then the full phrase (for example, *bisector*, *perpendicular*, and then *perpendicular bisector*) to help them see the broad category as well as the specific meaning within it.

Visit ***go.SolutionTree.com/MathematicsatWork*** *for a free reproducible version of this table.*

In addition to the suggestions in table 3.1, figure 3.2 (page 40) presents several structured mathematical language routines you can use to support and ensure student engagement and collaboration in the vocabulary for the lesson.

Direct vocabulary instruction can improve mathematical understanding as well. Irene T. Miura and Jennifer M. Yamagishi (2002) find that direct instruction in language, symbols, and their connections can help your students acquire concepts.

It can take up to twelve experiences for a word to have meaning for a student (McEwan-Adkins, 2010). This is why students need to daily be using the language of mathematics and tools like graphic organizers and active word walls, where students receive time for their own sense making and reasoning about the words or symbols. These are important tools for continued use of vocabulary words and mathematics notation in classroom discussions and lessons.

In addition to the mathematical language routines in figure 3.2, figure 3.3 (page 41) presents several graphic organizers you can use to support and ensure student engagement in the vocabulary for the lesson.

These mathematical language routines and graphic organizers provide a sampling of ideas you could use to embed vocabulary instruction purposefully into your lesson plans.

Mathematical Language Routines
Directions: Of all the vocabulary strategies listed, which types do you currently use during your mathematics lessons as part of instruction? Which might you try?
Cloze Passages Use cloze passages for steps to build student familiarity with text structures, vocabulary, and comprehension. 1. Retype a passage and place a blank line in place of strategic words. 2. Have students read the passage and try to determine from context the words that might fit in the blanks. Students earn five points for using words from the textbook and three points for using another word that makes sense to them in the context of the passage. Provide students a word bank as a scaffold if needed. 3. Have students share their passage with a partner and provide feedback to their peers. Discuss whether words students chose that differ from the author's make sense in the context of the passage. 4. Share the final answers, and have students celebrate the words they already know and make index cards for those words they are working toward learning.
Survival Words This six-step activity uses inquiry to determine how well students may understand vocabulary words they need to successfully read a mathematical task. 1. Choose several words that students may get tripped up by and that they are likely to see again, or use a list of high-frequency words from the mathematics unit the class generates. 2. Have students make a chart that has the following six column headings: Word, A, B, C, D, and Meaning. 3. Have students copy each word down in the first column of the chart and check the appropriate A, B, C, or D category for each word. A. I know the meaning, and I use the word. B. I know the meaning, but I don't use the word. C. I've seen the word before, but I don't really know it. D. I've never seen the word before. 4. Ask students to write the meanings of as many of the words as they know in the Meaning column. 5. Break students into groups to share the words and meanings they are most confident about. 6. Go over the charts with students, and answer any questions, giving additional information or clarification as they need it.
Three Reads The three reads protocol includes reading a mathematics scenario three times with a different goal each time. 1. **First read:** Read to understand the story—What is the problem about? If possible, remove the numbers and the question to focus students' thinking on what is happening in the story or task to reinforce sense making. 2. **Second read:** Identify the question and read to understand the mathematics. Add the numbers back into the problem and analyze the language used—What information is given? What don't you understand? What model can be created to represent the story? 3. **Third read:** Read the entire problem. Plan how to solve it—How can you solve the problem?
Highlights This four-step mathematical language routine encourages students to read word problems thoughtfully and document what the problem is asking. 1. Have students read directions to a problem set or a word problem. 2. Have students underline in one highlighter color what the problem is asking. *What is the question that the student is trying to answer?* 3. Have students underline the important information needed in a different highlighter color. *What information is needed to solve the problem?* 4. Discuss how problems are similar or different and help students use the vocabulary to understand the type of problem that is being asked.
Making Meaning This five-step activity is especially effective with multilingual learners, but it also helps any student make meaning of vocabulary. 1. Divide a 3- × 5-inch note card into quadrants. 2. In the upper-left quadrant, ask the student to write the vocabulary word in English or their first language.

3. In the upper-right quadrant, ask the student to write a synonym for the vocabulary word or to write the word in English if the upper-left quadrant has their first language.
4. In the lower-left quadrant, ask the student to define the word in any language using their own words, not a verbatim definition from a text.
5. In the lower-right quadrant, ask the student to make meaning of the word with a picture or example.

Word Wall

In this activity, the word wall displays words used throughout a unit. Students give meaning to the words on display.

1. Identify common unit vocabulary words.
2. List the words on chart paper or on a whiteboard in the classroom. Keep the word wall in the same part of the classroom all year.
3. As words from the unit appear during instruction, students write the meanings of the words and write an example or draw a picture in their notes or mathematics journal to make sense of the word. One student documents their work on the class chart paper, which can be made into anchor charts.

Tip: Reference the word wall routinely in class.

At the beginning of the unit, use a graphic organizer like a KIP chart that follows to help students make meaning of the words.

At the middle and end of the unit, use a graphic organizer like a foldable (see figure 3.3 for examples) to help students connect the words and show relationships.

Figure 3.2: Mathematical language routines.

Visit ***go.SolutionTree.com/MathematicsatWork*** *for a free reproducible version of this figure.*

Graphic Organizers

KIP

KIP is a graphic organizer that uses a chart to document important vocabulary: key vocabulary, information about the vocabulary, and a picture drawing of the word. The five steps are the following.

1. Create a chart like the one that follows.

Key	Information	Picture
Example:		

2. Students document the key vocabulary (K) in the chart. Each word will have its own chart.
3. Students write information (I) about the word—their own definition for the word.
4. Students draw a picture (P), if appropriate, for the word.
5. Students write or draw an example showing the word.

Frayer Model

This is a graphic organizer for vocabulary.

1. Have students write the word they are going to define in the middle of the graphic organizer.
2. In the upper-left corner, have students write the definition of the word (in their own words).
3. In the upper-right corner, have students write the facts or characteristics they know about the word.
4. In the lower-left corner, have students write or draw an example of the word.
5. In the lower-right corner, have students write or draw a nonexample of the word.

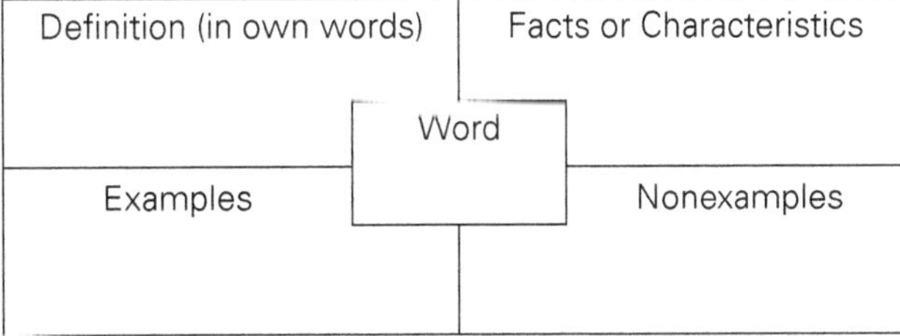

Figure 3.3: Mathematical language graphic organizers.

continued →

Foldable

A foldable is a tool used to organize vocabulary for any unit. Most foldables require one sheet of paper that you fold or cut to creatively organize words, definitions, and examples. You can use foldables to integrate reading, writing, thinking, data organizing, researching, and other communication skills into an interdisciplinary mathematics curriculum (Zike, 2003). Visit the *Math Equals Love* blog (https://mathequalslove.net/category/inbs/foldables/page/12) for more information about using a foldable.

Word:	**Definition:**	**Example:**
Perimeter	The sum of the side lengths of a polygon	7, 3, 3, 7 7+7+3+3=20

Source: Dale & O'Rourke, 1986; Frayer, Fredrick, & Klausmeier, 1969; Zike, 2003.

Visit ***go.SolutionTree.com/MathematicsatWork*** *for a free reproducible version of this figure.*

Figure 3.4 applies the activities and organizers from figures 3.2 and 3.3 to sample formal unit standards from grade 2, grade 4, middle school, and high school to show what strategies a team might choose for a specific unit standard. The tool in figure 3.4 is one that you and your team can use as you plan for vocabulary instruction within your units.

In order to use figure 3.4 as a team planning tool, you could list all the formal unit standards for one unit in the left column, and then specifically note the vocabulary (terms and notations) for the unit in the middle column and identify which language routines best support student understanding and agency in the right column.

Formal Unit Standard	Vocabulary	Mathematical Language Routine
Grade 2 Read standard three-digit numerals, write in word and expanded forms, and identify the place value of each digit, with and without models.	Place value Hundreds Tens Ones Standard form Word form Expanded form	Second grade is a critical year for developing base-ten number sense and place value understanding. Counting routines that include using a place value chart and base-ten tools give students an opportunity to see the vocabulary and values in a concrete way. Students can be involved through choral counting as a whole class or take turns by counting around the room as base-ten blocks are added to the place value chart. Folded sentence strips can be used to show the digits in standard form and stretched out to model expanded form. Incorporating words, numbers, visual models, oral counting, and base-ten tools allows for students to get a complete picture and makes meaning of the vocabulary that will shape their number sense understanding.
Grade 4 Explain why a fraction $\frac{a}{b}$ is equivalent to a fraction $\frac{(n \times a)}{(n \times b)}$ by using visual fraction models, with attention to how the number and size of the parts differ even though the two fractions themselves are the same size. Use this principle to recognize and generate equivalent fractions.	Fraction Equivalent fraction Fraction model Numerator Denominator	A graphic organizer with visuals would be extremely helpful for students in the fraction unit. The KIP model, the Frayer model, and a foldable would all be appropriate strategies to help students organize the words and give them time to make their own meaning. The term *fraction* was previously introduced in grade 3; however, it is important that students understand not just the word, but also how to represent a fraction visually and be able to provide multiple examples.

Middle School Understand the concept of a unit rate $\frac{a}{b}$ associated with a ratio $a:b$ with $b \neq 0$, and use rate language in the context of a ratio relationship.	Ratio Rate Unit rate	These words are new for students in sixth grade, so it is important to spend time guiding them through different activities to make meaning of the words in isolation and within the context of real-world problems. The KIP model and the Frayer model would be appropriate strategies to help students organize the words and give them time to make their own meaning. A word wall would also help keep the terms visible for students throughout the unit and for the whole year.
High School Solve linear equations and inequalities in one variable, including equations with coefficients represented by letters.	Coefficient Terms Factors Variable Expression Linear Equation Inequalities	Even though some of these words aren't new for students in high school, a foldable would be an appropriate graphic organizer to help make sense of the language. In addition, students could use it as a tool to reference when working on tasks. The words *expression*, *variable*, and *terms* might have different meanings for students outside of mathematics, so it is important to point out the differences. When students are solving contextual linear equations and inequalities in one-variable problems, a three reads strategy would support the students with slowing down the reading and making sense of what the problem is asking.

Figure 3.4: Teacher team discussion tool—Planning for mathematical language routine.

Visit ***go.SolutionTree.com/MathematicsatWork*** *for a free reproducible version of this figure.*

The goal of the mathematical language routines is to provide opportunities for students to create their own meaning so they can confidently use the words they learn in written and oral communication. As you include mathematical language routines in your lesson design, you will need to formally assess vocabulary words and notations on your unit assessments. Students need to be able to read the words and understand them, hear the words and understand them, and use the words in their explanations and work.

Additionally, ask your students to read equations and expressions with notations to ensure they can read and use the notations during the whole-group or small-group discourse of the lesson. For example, $3 < 5$ should be read as "three is less than five." Too often, students think an inequality is "more" because it is taught as an alligator eating more or they have not yet learned how to read the symbol correctly. In kindergarten, $3 + 5 = 8$ is read as "three plus five is the same as eight." In middle school geometry notations, $m{<}A = 30°$ is stated as "the measure of angle A is equal to 30 degrees." In high school, $\sin x = 30$ is read as "the sine of x is equal to 30" with the 30 as an implied degree measure. Mathematical language, both words and notations, is crucial to students' being able to listen, talk, and learn from one another through discourse and peer review.

You and your teacher team can also identify daily language objectives to support language development for multilingual learners. In addition to the mathematical language you and your team agree on, the language objectives identify the reading, writing, speaking, or listening skill that may be used to support language awareness (Echevarria, Vogt, & Short, 2010).

Language objectives specifically outline the type of language skill that students will need to learn and use to meet the learning outcomes of a lesson. For example, in figure 3.4, the language objective for the grade 4 essential standard could be, *Students can use the mathematics vocabulary to explain why the fractions are equivalent.* By identifying the specific language skill in the lesson, you and your team can use language objectives as a tool for incorporating the language skill into your intentional lesson design (Hansen-Thomas, Langman, & Farias, 2018).

Recall that part of the purpose of your work with implementing these lesson-design criteria is answering the third critical question of a PLC: How will we respond in class when some students do not learn? Sometimes, that response must be to develop the intentional plan for students to process understanding of the mathematical language (vocabulary and mathematics notations) being used by creating multiple opportunities for student demonstrations of those meanings.

Reflection on Practice

Up to this point, the lesson-design elements have helped you to use the essential learning standards for the unit, identify the learning target for the lesson, determine why it is so important for students to learn, create a context for learning via the prior-knowledge warm-up task you choose, and finally consider the mathematical language in the unit that can be a barrier to student meaning making and learning.

Equipped with this wisdom, you and your colleagues now make the most important and powerful decision in your daily professional work as a teacher of mathematics: you *choose* the mathematical tasks you will use to help your students demonstrate evidence of learning the new learning target before the lesson ends.

This is a lot of responsibility. Expert teachers ensure students will experience both higher-level-cognitive-demand tasks and lower-level-cognitive-demand tasks that align to the learning target for that day.

TEAM RECOMMENDATION

Use Mathematical Language Routines During Instruction

Consider the following recommendations with your team.

- As a grade-level or course-based team, select the mathematical language (including notations) in advance of the unit, and post or give the language to students on the day the unit starts.
- As you teach the unit, be sure to refer to the words, adding definitions, giving context, and allowing time for students to make meaning.
- Through the use of the mathematical tasks you choose and small-group discourse, assess for student understanding of the vocabulary, while also monitoring student misconceptions.
- As you teach, use mathematical language routines or graphic organizers, such as the ones figures 3.2 (page 40) and 3.3 (page 41) describe, to help clarify the meaning of words and support long-term retention of vocabulary.

Visit **go.SolutionTree.com/MathematicsatWork** for free reproducible versions of tools and protocols that appear in this book, as well as additional online only materials.

CHAPTER 4

A Balance of Mathematical Tasks

> What should be the nature of mathematics that students learn—facts, skills and procedures or concepts and understanding? How should students learn mathematics—teacher directed with a focus on memorization, or student discovery through reasoning and discovery?
>
> —*Phillip S. Jones and Arthur F. Coxford Jr.*

Although hard to believe, the preceding quote in the epigraph from Jones and Coxford was written in 1970. The debate between procedural knowledge and conceptual understanding in mathematics goes back a long time. Matthew R. Larson and Timothy D. Kanold (2016) establish that the debate surrounding the level of cognitive demand of the mathematical tasks you choose to teach lessons each day began in the 1820s and still exists today.

Furthermore, when it comes to mathematics task design, Larson and Kanold (2016) describe an equilibrium approach that balances the emphasis procedures and conceptual understanding as the best way forward for mathematics education. NCTM (2014) states that students need sufficient time in daily lessons to engage in tasks that promote problem solving and reasoning to make sense of new mathematical ideas.

As described so far, lesson design begins with the development of your team's understanding of the essential learning standards for the unit and the daily learning targets and essential questions for each lesson in the unit. You also work to identify prior-knowledge standards and the mathematical language necessary to help your students communicate their learning and persevere throughout the lesson. These lesson-design and planning actions help you to answer the first critical question of a PLC for collaborative teams: What do we want all students to know and be able to do?

And now, you choose the *mathematical tasks* you and your colleagues will use each day. Few other decisions you make on a daily basis have the same strength of impact on student learning as the choice of tasks you use for your lesson. The mathematical tasks and activities you choose for each lesson of the unit will help you to answer the second critical question of a PLC for collaborative teams: How will we know if students learn it?

You choose the mathematical tasks for the lesson based on your judgment that those tasks will help your students demonstrate an understanding of the learning target for the day. These task choices will also impact the rigor of the student learning experience. From an equity perspective, you want to select tasks characterized as low-threshold, high-ceiling tasks (NRICH Team, 2019) that provide access and potential scaffolding entry points for all students.

Meaningful and Relevant Mathematics

At the same time, the task you choose should have the potential to engage students in challenging mathematics and to help students understand the relevance of learning the essential standard and creating meaning and thinking at a much deeper reasoning level (Smith et al., 2017). *Relevance* and *meaningfulness* become important terms to consider as you choose the tasks in the design of your daily mathematics lessons.

TEACHER *Reflection* 

Take a moment to reflect on and discuss with a colleague the difference between *relevant* mathematics and *meaningful* mathematics.

Although some educators use the words *relevant* and *meaningful* interchangeably, these words support significantly different ideals for mathematics lesson design.

Think of relevant mathematics as mathematics that contains "the important stuff"—that which establishes a context for the learning standard of the lesson discussed in chapter 1 (page 11). *Relevant* mathematics represents mathematics units of study that contain essential mathematics and mathematical tasks students *need to know*. To determine if a lesson is relevant, ask, "Does the lesson present essential mathematics? Why must students learn this mathematics lesson today?" Think of relevant mathematics as establishing a *context* for learning the mathematics standard of the daily lesson. Meaningful mathematics, on the other hand, presents a different type of lesson-design challenge.

Think of meaningful mathematics as an aspect of your lesson design that is all about the student's point of view and engagement during the lesson. *Meaningful* mathematics contains elements that create student agency in learning through reasoning and sense making, while also connecting to students' prior knowledge (see chapter 2, page 23) and understanding. To determine if a lesson is meaningful, ask, "How does the daily lesson design create meaning for students? What will students be *doing* during each part of the lesson (see chapter 5, page 57)? And will students persevere when they get stuck?"

Meaningful mathematics is grounded in your daily choice of mathematical tasks and ways in which the tasks draw on multiple sources of knowledge from the student. This includes intentionally tapping into students' prior knowledge and experiences, including their cultural, linguistic, family, and community resources, as you teach. This is one of five equity-based teaching practices Julia Aguirre, Karen Mayfield-Ingram, and Danny Bernard Martin (2013) offer. When teachers draw on these resources to make mathematics meaningful for students, they help "students bridge everyday experiences to learn mathematics" (Aguirre et al., 2013, p. 43). Students need to see themselves in the mathematics and use the mathematics to help them make sense of the world.

When you and your team acknowledge students' cultural, linguistic, family, and community experiences, you are positioning students as doers of meaningful mathematics. *NCSM Essential Actions: Framework for Leadership in Mathematics Education* (NCSM, 2020) states that you and your team need to prioritize building cultural awareness because students need access to meaningful and relevant mathematics. Mathematics teachers know that every student brings a unique background and different cultural experience into the classroom. Consider the following reflection questions from NCSM (2022):

- How do we honor the different cultures that are represented in our classrooms and systems?
- How do we intentionally embrace the diversity of the cultures and experiences in order to create a culture of belonging?
- How can we shape our school community so that students see themselves in the classroom and in the mathematics tasks assigned?
- How can diversity be embraced to serve as a resource to make instruction more accessible for all learners? (p. 19)

Your team moves toward a more equitable instructional process when you engage in systemic reflection on the cultures and traditions within the classroom. This ensures that you and your team are focused on meaningful and relevant mathematics activities for every student.

As you understand that tasks form the basis for students' opportunities to learn what mathematics is and how one *does* mathematics, it creates a realization that your task selection for the lesson is of high priority and importance. The level and kind of thinking required by the mathematics instructional tasks you choose ultimately influence what your students learn that day.

Essentially, low-threshold, high-ceiling mathematics tasks present activities where everyone in the student group can begin and work at their own level, yet the tasks also offer possibilities for learners or teams of learners to do much more challenging mathematics using the tasks as well.

According to Melissa D. Boston and Margaret S. Smith (2009), a *mathematical task* is a single complex problem or a set of problems that focuses students' attention on a specific mathematical idea. Mathematical tasks include activities, examples, or problems that students complete as a whole class, in small groups, or individually. The tasks provide the rigor (levels of complex reasoning) you choose to determine the pathway of student learning, and to assess student success along that pathway during the lesson.

Additionally, a growing body of research links students' engagement in higher-level-cognitive-demand tasks to overall increases in mathematics learning, not just in the ability to solve problems (Hattie, 2012; Resnick, 2006; Smith et al., 2017).

Balanced Mathematical Task Routines

There are several ways to label the cognitive demand or rigor of a mathematical task; however, for the purposes of this book and series, we classify tasks as either lower-level cognitive demand or higher-level cognitive demand as Margaret Schwan Smith and Mary Kay Stein (1998) define in their task analysis guide that is printed in full as an appendix (page 115). Their seminal research work shines a bright light on one of the most important decisions we make each day: the nature of the mathematical tasks chosen to help students learn a standard.

Lower-level-cognitive-demand tasks typically focus on memorization or rote procedures without attention to the properties that support those procedures (Smith & Stein, 2011). *Higher-level-cognitive-demand tasks* are tasks for which students do not have a set of predetermined procedures to follow to reach resolution, or, if the tasks involve procedures, they require that students justify why and how they perform the procedures.

Examples of lower- and higher-level-cognitive-demand tasks (for use either in class, for homework, or on a mathematics assessment) for various grade levels appear in figure 4.1.

<table>
<tr><td colspan="2">Directions: Choose the most appropriate grade level that follows, and as a collaborative team, discuss why each of the questions meets the cognitive-demand levels to which it's assigned using the descriptions for lower- and higher-level-cognitive-demand tasks in the appendix (page 115).</td></tr>
<tr><td colspan="2">Grade 1: I can solve addition and subtraction word problems up to 20.</td></tr>
<tr><td>Lower-level-cognitive-demand task:
How many flowers are in the garden?
</td><td>Higher-level-cognitive-demand task:
We can have 12 flowers in the garden below. Make different combinations of red and yellow flowers in the garden. Did you find all the combinations? How do you know?
</td></tr>
<tr><td colspan="2">Grade 3: I can interpret whole-number quotients using objects.</td></tr>
<tr><td>Lower-level-cognitive-demand task:
Divide into groups of three.
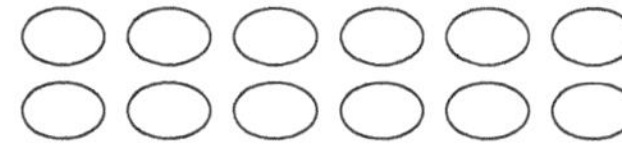</td><td>Higher-level-cognitive-demand task:
Kyle sold some tubs of peanut butter cookie dough. Each tub costs $8. Kyle collected $32. How many tubs of peanut butter cookie dough did Kyle sell? Show your work and write an equation.
Explain how you solved the problem.</td></tr>
<tr><td colspan="2">Grade 6: I can write an inequality of the form $x < c$ or $x > c$.</td></tr>
<tr><td>Lower-Level-Cognitive-Demand Task:
Graph the inequality.
$x \leq 14$</td><td>Higher-Level-Cognitive-Demand Task:
Tony graphed $x < -4$ below. Is he right? Explain how you know. If he is wrong, correct his error.

</td></tr>
<tr><td colspan="2">High school (algebra 2): I can simplify or solve rational expressions and equations.</td></tr>
<tr><td>Lower-level-cognitive-demand task:
Solve for x. Identify extraneous solutions.
$\frac{2}{x} = \frac{x}{x^2-8}$</td><td>Higher-level-cognitive-demand task:
If $\frac{2}{a-1} = \frac{4}{y}$ where $y \neq 0$ and $a \neq 1$, what is y in terms of a?
For what values of a will y be negative? Show your work and justify your reasoning.</td></tr>
</table>

Source: High school example adapted from the College Board, n.d.

Figure 4.1: Examples of lower- and higher-level-cognitive-demand mathematical tasks.

When planning each unit together, your team should determine lower-level-cognitive-demand tasks and higher-level-cognitive-demand tasks for each essential learning standard (see the *Mathematics Unit Planning in a PLC at Work* series). The agreed-on tasks provide examples of the types of tasks you need to choose and use during your daily lessons.

A key word regarding mathematical task rigor is *balance*. Your choice of daily mathematical tasks for teaching the lesson should reveal a balance of procedural fluency and conceptual understanding proficiency throughout the lesson. Generally, the rigor-balance ratio is recommended to be about 50–50 (higher- to lower-level cognitive demand) during any unit, with the procedural knowledge and rote memorization student tasks *following* the development of conceptual understanding of the standards. Procedural fluency tasks should be built on a foundation of conceptual understanding task development during the lesson (NCTM, 2014).

Why Balancing Mathematical Tasks Is Important

In order for students to develop true understanding of the essential learning standards, procedural fluency *and* conceptual understanding should be used in tandem:

> Procedural fluency and conceptual understanding are often seen as competing for attention in school mathematics. But pitting skill against understanding creates a false dichotomy. . . . Understanding makes learning skills easier, less susceptible to common errors, and less prone to forgetting. (Kilpatrick et al., 2001, p. 122)

Mathematics instruction should focus on developing skills and understanding for each student, as well as the ability for each student to reason with their understanding and skills to solve problems. Skills and understanding support each other, and are both necessary to be an effective problem solver (Larson & Kanold, 2016).

Calls for education to change based on the economy's need for deeper learning and less rote learning, first pointed out by James W. Pellegrino and Margaret L. Hilton (2012) in their landmark report, have only intensified in subsequent years (Bakhshi, Downing, Osborne, & Schneider, 2017; National Academies of Sciences, Engineering, and Medicine, 2017; Rainie & Anderson, 2017; World Economic Forum, 2016). In the following personal story, Sarah Schuhl references a conversation with a teacher on a chosen daily lesson-planning technique (or lack of one).

In *Balancing the Equation*, authors Larson and Kanold (2016) state, "A modern definition of *mathematical literacy* includes student development of skills and procedures, conceptual understanding, problem solving, and a disposition to expend effort and persevere when learning mathematics and solving problems" (p. 62). This explanation establishes a working definition for the rigor you should seek in every mathematics lesson.

Personal Story  SARAH SCHUHL

A teacher once told me the reason he loves teaching mathematics is because all he has to do is open the book each day to the lesson and show the examples from the text to students and assign homework. As I thought about his lesson-planning technique, I realized that he had a major misconception about teacher resources. How did he verify that the tasks in the publisher resources fully met the intent of the essential learning standard? Often the tasks in curriculum resources are in isolation from other lessons, focused only on the written learning target for that day's lesson, and they may or may not be a balance of lower- and higher-level tasks. In contrast, a teacher may need to combine content ideas, include specific process standards, or carefully choose which tasks to use because there may not be enough time to share and engage students in each task every day. In general, teams should ask, What is the purpose for using each mathematical task related to student learning during the lesson? What is the plan? Why should we use each task, and is there a balance in the cognitive-demand level of the tasks we choose?

For example, it benefits students to use objects, number lines, skip counting, groups, and area models to build an understanding of multiplication before they can develop procedural fluency. Students need strategies to learn the multiplication facts beyond pure memorization with flash cards or timed tests. Procedural fluency is not just how fast a student can answer a question. Procedural fluency occurs when students are efficient, accurate, and flexible as thinkers (Bay-Williams & Stokes-Levine, 2017). Understanding of and development of flexible thinking with numbers are essential for elementary school students.

To select balanced mathematical tasks you use each day, start with sources like publisher materials or lessons you have previously used. From there, you may search websites for mathematical tasks and visit **go.SolutionTree.com/MathematicsatWork** or scan the QR code in this book to access a set of free online resources, or look at tasks in other textbooks and resources as well.

Thus, your teacher team should engage in meaningful dialogue and discussion about the choices of mathematical tasks members use to teach each lesson of the unit. You can use figure 4.2 (page 50) to engage your team in reflective conversations about how and why tasks are chosen to teach the lesson.

Not all tasks need to be common across the team for each and every lesson within a unit; however, there needs to be consistency in the interpretation of the intended rigor of the essential learning standards and the expectations for proficiency. Even though the specific tasks you and your colleagues choose may not match exactly, the level of rigor students experience from one class to the next should be consistent. Also, the types of strategies and tools (manipulatives and others) students might utilize throughout the unit should be consistent. This ensures each and every student has access to rigorous learning opportunities.

As a team, identify and commit to the use of higher- and lower-level-cognitive-demand tasks on a consistent basis within your instruction. The balance of the types of tasks you choose supports student learning, encourages perseverance, and provides opportunities for you to truly assess students and respond to specific student needs.

TEAM RECOMMENDATION

Examine Your Cognitive-Demand Task Balance

- Mathematical tasks can be a single problem or a set of problems and activities you use to develop student understanding for each essential learning standard of the unit.
- It's important to use a balance of lower- and higher-level-cognitive-demand tasks within the daily lessons of the unit in order to support a balance of student development in procedural fluency and conceptual understanding.
- Higher-level-cognitive-demand tasks support increased student exploration, communication, and reasoning and allow for multiple solution pathways or representations.

How you select tasks and require students to engage in them matters. This selection establishes the expectation level for instruction and ensures that you engage each student in higher-level-cognitive-demand tasks, providing the opportunity for students to go deep with mathematics, one of the equity-based mathematics teaching practices (Aguirre et al., 2013).

Once you and your team discuss the lower-level-cognitive-demand tasks and higher-level-cognitive-demand tasks, you will then have to plan how to implement the tasks. Selecting a balance of cognitive-demand tasks is a starting point, and your next team action is to discuss the implementation of the mathematical tasks to maintain the mathematical tasks' rigor.

How to Implement Balanced Mathematical Task Routines

Part of the struggle with implementing higher-level-cognitive-demand tasks is maintaining student perseverance and rigor throughout the task. Lower-level tasks are typically easier to implement, as they require less cognitive demand for the student. The solution pathway is predictable and requires less thinking.

Directions: As a team, use the following questions to discuss how you currently select higher- and lower-level-cognitive-demand tasks within your lesson-design process.

1. Describe some of your favorite mathematical tasks to use during this unit and how you use them to teach the corresponding essential learning standard.

2. How do you define and differentiate between higher-level-cognitive-demand and lower-level-cognitive-demand tasks for each essential learning standard of the unit?

3. What percentage of your current mathematics tasks that you use during the unit of instruction fall into the lower-level-cognitive-demand category, and what percentage fall into the higher-level-cognitive-demand category? (Provide an average.)

4. How do you work as a team to select specific common higher-level-cognitive-demand and lower-level-cognitive-demand mathematics tasks that all students of the grade level or course will experience for each essential standard of the unit?

5. How will the tasks your team uses support students' move from conceptual understanding learning to procedural fluency learning?

Figure 4.2: Teacher team discussion tool—Choosing mathematical tasks for lesson design during the unit.

Visit ***go.SolutionTree.com/MathematicsatWork*** *for a free reproducible version of this figure.*

However, implementing higher-level-cognitive-demand tasks requires students to make sense of the problem, figure out an entry point into the task, make connections, draw conclusions, and then verify and communicate their thinking.

Have you ever had students shut down when you try something new in class or when you give them a problem they perceive as too hard for them? If you have, then you know it is not fun when it happens. In relation to balancing mathematical tasks, you and your team can examine instructional routines during a lesson that maintain student perseverance as well as the rigor of the mathematical tasks.

Take a moment to consider how you currently implement a balance of higher- and lower-level-cognitive-demand tasks.

TEACHER *Reflection* 

What might you learn about your students' understanding of the essential learning standard depending on the cognitive demand of the mathematical tasks you use during instruction?

How do you use higher-level-cognitive-demand tasks to provide feedback to individual students and groups of students during the lesson?

In order to support student perseverance through the higher-level-cognitive-demand mathematical tasks, you and your team members teach students how to act and respond like mathematicians. Does that surprise you?

Mathematicians are able to communicate their reasoning using accurate vocabulary. Mathematicians understand and see the importance in the process of learning and are able to reason through their own ideas and learn from early mistakes or errors while listening to and learning from others.

Mathematicians are able to use and connect multiple representations and solve problems in many ways. When students learn to represent, discuss, and make connections among mathematical ideas in multiple forms, they demonstrate deeper mathematical understanding and enhanced problem-solving abilities (Donovan & Bransford, 2005; Janvier, 1987).

To facilitate student perseverance during the lesson, there are five practices you can use to maintain student engagement and effort.

1. Connect the mathematical tasks to the *essential learning standard* for the unit and the daily learning target for the lesson.
2. Present mathematical tasks during the lesson that are of *higher-level cognitive demand.*
3. *Monitor* students' *initial responses* to the mathematical tasks through observation.
4. Ensure the mathematical tasks expect a process of student *reflection* and *communication with peers.*
5. Expect students to *compare and contrast solution pathways* to the mathematical tasks.

In order to engage in these five perseverance practices, it is necessary to be clear on the essential learning standard or standards for the lesson and be sure you and your team choose tasks that will support student engagement in that standard. (You will explore more about student engagement in the next chapter, page 57.)

Thus, every day, you make a clear communication of the standard (an *I can* statement or an essential question) to the students and connect it directly to the mathematical tasks you choose for the lesson.

Consider the five higher-level-cognitive-demand tasks in figures 4.3 (page 52), 4.4 (page 52), 4.5 (page 53), 4.6 (page 53), and 4.7 (page 54), representing kindergarten; grades 2, 4, and 7; and high school. A critical first step in selecting and planning to use a higher-level-cognitive-demand task in class occurs when your team works through the solution pathways for the mathematical task together and discusses the nuances of the problem. You do this team activity *before* you use the mathematical task with your students.

Your collaborative team can use the sample tasks from figures 4.3–4.7 to explore elements of planning for the formative assessment process during the lesson. As you use higher-level-cognitive-demand mathematical tasks within the context of what students are to *say and do*, your plans for creating a formative assessment process in class can unfold. You will explore more about the formative process in chapter 5 (page 57). Visit **go.SolutionTree.com/MathematicsatWork** or scan the QR code in this book for additional grade-level sample tasks similar to figures 4.3–4.7 and a blank template.

Kindergarten **Standard:** Represent addition and subtraction with objects, fingers, mental images, drawings, sounds (for example, claps), acted-out situations, expressions, or equations.	
Mathematical task: Read the problem to the student—Julia has 9 cupcakes. She shares 4 cupcakes with her friends. How many cupcakes does Julia have now? Show your thinking with objects, words, pictures, or numbers.	
What types of misconceptions do you anticipate students will struggle with during the task?	
What scaffolding questions can you ask students to help guide their work?	
What are the ways feedback for the student solution pathway or explanation can be provided during the lesson (teacher to student, student to student)?	
What will happen if students finish early?	

Figure 4.3: Planning for the formative assessment process—Kindergarten.

Visit ***go.SolutionTree.com/MathematicsatWork*** *for a free reproducible version of this figure.*

Grade 2 **Standard:** Explain why addition and subtraction strategies work, using place value and the properties of operations.	
Mathematical task: Adam bought 17 tickets at the fair. His brother gave him 9 more. Chelsea had 9 tickets from last year, and she bought 17 tickets when she got to the fair. Do Adam and Chelsea have the same number of tickets? How do you know? Explain using pictures, numbers, and/or words.	
What types of misconceptions do you anticipate students will struggle with during the task?	
What scaffolding questions can you ask students to help guide their work?	
What are the ways feedback for the student solution pathway or explanation can be provided during the lesson (teacher to student, student to student)?	
What will happen if students finish early?	

Figure 4.4: Planning for the formative assessment process—Grade 2.

Visit ***go.SolutionTree.com/MathematicsatWork*** *for a free reproducible version of this figure.*

<table>
<tr><td colspan="2">Grade 4
Standard: Understand a fraction $\frac{a}{b}$ with $a > 1$ as a sum of fractions $\frac{1}{b}$.</td></tr>
<tr><td colspan="2">Mathematical task: Decompose the fraction $\frac{9}{12}$ in two different ways. Show your work.
Be sure to explain how you decomposed your fraction.</td></tr>
<tr><td>What types of misconceptions do you anticipate students will struggle with during the task?</td><td></td></tr>
<tr><td>What scaffolding questions can you ask students to help guide their work?</td><td></td></tr>
<tr><td>What are the ways feedback for the student solution pathway or explanation can be provided during the lesson (teacher to student, student to student)?</td><td></td></tr>
<tr><td>What will happen if students finish early?</td><td></td></tr>
</table>

Figure 4.5: Planning for the formative assessment process—Grade 4.

Visit ***go.SolutionTree.com/MathematicsatWork*** *for a free reproducible version of this figure.*

<table>
<tr><td colspan="2">Grade 7
Standard:
1. Compute unit rates associated with ratios of fractions, including ratios of lengths, areas, and other quantities measured in like or different units.
2. Recognize and represent proportional relationships between quantities.</td></tr>
<tr><td colspan="2">Mathematical task: Family Yogurt sells frozen yogurt at a price based on the total weight of the yogurt and the toppings in a cup. Each member of Gia's family weighs their cup and pays the corresponding amount shown in the following chart.
<table><tr><td>Cost ($)</td><td>5</td><td>4</td><td>2</td><td>3.20</td></tr><tr><td>Weight (oz)</td><td>12.5</td><td>10</td><td>5</td><td>8</td></tr></table>
Does everyone pay the same cost per ounce? Show your work and explain how you know.
Is the cost proportional to the weight? Explain your answer.</td></tr>
<tr><td>What types of misconceptions do you anticipate students will struggle with during the task?</td><td></td></tr>
<tr><td>What scaffolding questions can you ask students to help guide their work?</td><td></td></tr>
<tr><td>What are the ways feedback for the student solution pathway or explanation can be provided during the lesson (teacher to student, student to student)?</td><td></td></tr>
<tr><td>What will happen if students finish early?</td><td></td></tr>
</table>

Figure 4.6: Planning for the formative assessment process—Grade 7.

Visit ***go.SolutionTree.com/MathematicsatWork*** *for a free reproducible version of this figure.*

High School **Standard:** Derive the equation of a circle of a given center and radius using the Pythagorean theorem; complete the square to find the center and radius of a circle given by an equation.	
Mathematical task: $x^2 + y^2 - 6x + 8y = 144$ The equation of a circle in the *xy*-plane is shown above. Determine the *diameter* of the circle using an algebraic solution pathway, and verify the solution is correct using a graphical representation. Be sure to show your work and explain your reasoning.	
What types of misconceptions do you anticipate students will struggle with during the task?	
What scaffolding questions can you ask students to help guide their work?	
What are the ways feedback for the student solution pathway or explanation can be provided during the lesson (teacher to student, student to student)?	
What will happen if students finish early?	

Figure 4.7: Planning for the formative assessment process—High school geometry.
*Visit **go.SolutionTree.com/MathematicsatWork** for a free reproducible version of this figure.*

Solving each task before the lesson builds understanding of what students have to know and be able to do, reveals where students may have misconceptions, and provides an opportunity to plan for active student engagement and learning. Solving the task with your team provides information about possible solution strategies or pathways students might demonstrate. You and your team can also share anticipated misconceptions to assist with developing questions and responses to address possible challenges. These discussions are needed to support differentiation and provide just-in-time support during Tier 1 (core) instruction, which is explored more in chapter 7 (page 93).

Additionally, your team must discuss how to use various technologies during the implementation of certain mathematical tasks. The use of technologies such as 1:1 notebooks and software such as ChatGPT-4 can expedite students' making connections between and among multiple representations. When utilizing technology, be sure to discuss how to best support meaningful student exploration and student-to-student discourse (see chapter 5, page 57).

Reflection on Practice

Higher-level-cognitive-demand mathematical tasks are those that provide "opportunities for students to explain, describe, justify, compare, or assess; to make decisions and choices; to plan and formulate questions; to exhibit creativity; and to work with more than one representation in a meaningful way" (Silver, 2010, p. 2). In contrast, lessons or tasks with low cognitive demand are "characterized as opportunities for students to demonstrate routine applications of known procedures or to work with a complex assembly of routine subtasks or non-mathematical activities" (Silver, 2010, p. 2).

So far, you have examined lesson-design elements to help you identify daily learning targets for each essential standard and determine why those standards are important, and you have created a context for learning via the prior-knowledge routine of your choice. You have also considered how the mathematical language in the unit can be a barrier to student understanding and learning, and you have discovered the expectation to use a balance of higher- and lower-level-cognitive-demand mathematical tasks throughout the unit. These lesson-design and planning actions serve the first two critical questions of a PLC (DuFour et al., 2016): What do we want all students to know and be able to do? and How will we know if students learn it?

In order to implement your chosen mathematical tasks for each lesson well, your lesson planning and design next need to carefully consider the type of discourse students will use to engage in each mathematical task. In the next chapter, you explore whole-group versus small-group discourse activities and how to balance the two types of student communication during the lesson.

TEAM RECOMMENDATION

Implement Balanced Mathematical Tasks

- Agree that balancing the cognitive demand and rigor of the mathematical tasks you choose is essential to student sense making, reasoning, and communication.
- Consistently plan for the formative feedback process by considering the types of questions used and the feedback provided during the tasks to sustain student perseverance.

Visit **go.SolutionTree.com/MathematicsatWork** for free reproducible versions of tools and protocols that appear in this book, as well as additional online only materials.

CHAPTER 5

Mathematical Discourse Routines

Tell me and I forget, teach me and I may remember, involve me and I learn.

—*Benjamin Franklin*

The nature of the classroom discourse during the lesson is an essential decision as you design how to engage students in the mathematical tasks of the lesson. The primary research-affirmed purpose of the mathematics lesson is to *actively engage all students* in demonstrating evidence of learning the essential learning standard and, more specifically, the learning target for that day. In order to honor this lesson-design purpose, you carefully consider the nature of the student discourse during the lesson. Student learning is enhanced when students have intentional opportunities to work with their peers, communicate what they are thinking, and learn to expand their understanding of the mathematical ideas (Van de Walle, Karp, & Bay-Williams, 2019).

The nature of the discourse design through student interactions has profound implications for how students view themselves as *learners of mathematics*. The student discourse should:

> help students see themselves as people who can know, do, and make sense of mathematics, challenging aspects of marginality that lie within students' own identities. The language choices that teachers and students make, the norms that are set in the classroom, and the types of mathematical interactions that are or are not encouraged can all support or inhibit the development of productive attitudes and practices toward mathematics. (Smith et al., 2017, p. 140)

Additionally, NCTM (1991) highlights the importance of engaging your students in peer-to-peer discourse *during* the lesson: "Teachers, through the ways in which they orchestrate discourse, convey messages about whose knowledge and ways of thinking and knowing are valued, who is considered able to contribute and who has status in the group" (p. 20).

Take a moment to complete the teacher reflection that follows. Use the teacher reflection to discuss current strengths with your mathematical discourse routines.

TEACHER *Reflection* 

Where and when are student peer-to-peer conversations currently taking place as part of your mathematics lessons?

Who is doing most of the talking throughout a lesson—you to your students, or your students with one another?

Intentional Use of Mathematical Discourse Routines

There are generally two primary types of mathematical discourse as you plan to actively engage your students in conversations with each task of the lesson.

1. **Whole-group discourse:** All students are learning together as the teacher facilitates discussion or models how to do a problem. The teacher calls on students throughout the lesson to respond to questions as the task solution unfolds, and reflects student responses back to other members of the class.
2. **Small-group discourse:** Students work together within a time limit and follow specific directions or prompts regarding how to participate in expected peer-to-peer shared discussions during core instruction. Students collaborate with other peers and proceed through strategies together for an assigned mathematics task or discussion prompt.

The manner in which you facilitate student discourse is essential in creating and supporting a classroom learning environment that values reasoning and sense making from the *student's point of view*. How you ask questions and implement discourse routines in your classroom has important implications for whether your instruction promotes a classroom culture that values students' voices, their experiences and ideas, and their cultures. The positioning of your students influences how students see themselves as members of a classroom community (Smith et al., 2017).

Margaret Schwan Smith, Michael D. Steele, and Mary Lynn Raith (2017) offer the following additional lesson-discourse reflection questions:

- Are all students' ideas and questions heard, valued, and pursued in the mathematics classroom?
- What mathematical ideas does the class examine and discuss?
- Whose thinking does the teacher select for further inquiry, and whose thinking does the teacher disregard during small-group and whole-class discussion?
- Who in the classroom is positioned as competent?
- Whose ideas are featured and privileged? (p. 95)

Through the use of either whole-group or small-group discourse, you decide what thinking to share and whose voices to hear. This has a profound impact on how knowledge is shared and created in your classroom, how deeply the task engages students, how they will receive feedback during the task, and who plays a primary role in that knowledge sharing and creation. In short, your discourse affects student agency toward learning the mathematics. Will the students persevere during the lesson and the tasks you have prepared? What if they get stuck while working on the tasks you have chosen? Then what? Thus, part of your lesson-design plan is to determine *how* your students will experience the process of learning the mathematical tasks you chose.

- Will students watch you model the task as you ask individual members of the class guiding questions from the front of the room? (Whole-group discourse)
- Will students discuss a strategy for working on the task with partners or in small groups as you circulate through the room and provide differentiated feedback and prompts for perseverance? (Small-group discourse)
- Or will students use some combination of both lesson communication styles?

As a teacher, you make these decisions on a daily basis in your lesson-design process. It is important to note that there are severe limitations to using whole-group discourse *only*.

As they relate to student perseverance in class, mind wandering and *cascading inattention*—overload in the mind—are due to the complexity of information you deliver to the students during the whole-group phase of a lesson. As John Hattie and Gregory Yates (2014) indicate, "student attention deteriorates over the course of a lesson" during whole-group discourse (p. 39). If students are listening to a teacher for over eight minutes, most likely, they are losing interest or no longer listening to the teacher speaking from the front of the room.

Thus, too much time spent using whole-group discourse instruction can be very damaging to student learning. The National Board for Professional Teaching Standards (2010), in its mathematics report, states it more succinctly: "In an environment of trust, students feel safe to communicate different points of view, to conduct open-ended explorations, to make mistakes, and to admit confusion or uncertainty in order to learn" (p. 36).

Hattie and Yates (2014) further indicate the effect size on student learning is significant (0.82 standard deviations above the norm) when students in your class see themselves as reliable and valuable resources for each other, and when they express their ideas, questions, insights, and difficulties *student to student.* Paul A. Kirschner and Carl Hendrick (2020) also share that discussion and collaboration are critical to support learning. They go on to state that one of the most effective learning techniques is when students are required to explain why a concept of fact is true.

Small-group discourse is all about what you *see* and *hear* the students thinking and reasoning about with one another. Part of your professional lesson-design responsibility is to ensure a daily balance of whole-group and small-group classroom discourse or discussions and be intentional about the instructional choices around the classroom discourse.

Remember, the intent of your lesson design is to be able to answer PLC critical questions 1–4 for your students during the lesson (DuFour et al., 2016).

1. What do we want all students to know and be able to do?
2. How will we know if students learn it?
3. How will we respond when some students do not learn?
4. How will we extend the learning for students who are already proficient?

To a great extent, if most of your lesson design is to use whole-group discourse from the front of the classroom, you will not be able to determine an answer to PLC critical question 2. You will only know the answer for the few students you call on.

On the other hand, student small-group discourse opens a window into student reasoning about mathematics. You *hear* and *see* the connections students make, the questions they ask, and the obstacles or misconceptions that can hinder their conceptual understanding.

It is mostly during the small-group discourse moments of the lesson that you can get a more accurate assessment concerning the essence of PLC critical question 2, How will we know if students are learning the standard? By listening carefully to students, you are in a better position to make decisions and implement strategies to support learning and push the level of reasoning and problem solving you expect higher.

The subsequent small-group student dialogue creates a community of learners who collectively build mathematical knowledge and proficiencies and *persevere* together. More importantly, you will improve your ability to effectively answer PLC critical question 3, How will we respond when some students do not learn (during small-group discourse activities; DuFour et al., 2016)? As your students make initial errors in their reasoning and thinking on the tasks you present to them, you are able to provide more accurate and meaningful student feedback to the lessons' mathematical tasks. Small-group discourse provides more accurate evidence of student thinking as you walk around the room during core instruction.

Thus, you deliberately structure opportunities for students to use and develop appropriate mathematical discourse as they reason and solve problems with one another. This means you give students opportunities to talk with one another, work together in solving problems, and discuss their mathematical thinking and understanding with their peers.

Structuring Whole-Group Discourse

Does this mean that whole-group discourse with the teacher lecturing or modeling from the front of the room is a bad practice? No. But it is currently an overused teacher practice that severely limits student learning if you rely on it too often and too narrowly. Your goal in planning is to seek *balanced student communication and discourse* around the tasks you choose for the lesson.

There are different situations where whole-group activities make sense. Three are as follows.

1. When you are introducing a new standard or modeling a standard with deeper connections and explanations, you most likely need to allow students to observe first, and then comment together and reflect on your modeling and thought process.
2. When you are leading the class through a mathematics task, you can occasionally blend in some student partner talk around a question or prompt arising from the nuances of the mathematical task you are asking students to think through. You can, as Mike Schmoker (2018) recommends, transform a traditional teacher-led lecture into an interactive lecture that raises student engagement. You do not really yield the whole-group discourse completely, as you

give your students brief moments of discussion time. You can keep the attention to the overall mathematics task flowing through your advice and modeling. This teacher action engages the students on a much deeper level than just calling on one or two students at a time during your whole-group moments of the lesson.

3. Sometimes, once you complete a small-group activity, it may help if you summarize or consolidate the important reveal from the small-group discourse learning experience, make mathematical connections students may have missed, or draw out the mathematical structure (Smith & Stein, 2011).

One caution, though, is to rarely go over with the whole group a problem the students just worked on during small-group discourse. If you don't heed this caution, students will learn quickly to not persevere on the task you are asking them to discuss with their peers, as they can just wait it out knowing you will show them how to do the mathematics task in a few minutes. Instead, during the summary of the learning after small-group discourse, have the students evaluate the thinking of the other student teams.

Structuring Small-Group Discourse

In order to balance whole-group and small-group discourse routines at different times within a lesson, your students should be working in groups or teams of students.

By setting up a culture of collaboration with your students, the transition between whole-group activities and small-group activities during the lesson becomes seamless. In order to facilitate both types of discourse well, it is necessary to use clear small-group and whole-group structures and expectations. You will explore how to use these structures for a more formative feedback purpose in the last part of this chapter. In the following personal story, Tim Kanold reflects on shifting to a classroom community of learners.

Personal Story  TIMOTHY KANOLD

By 1994, after eight years of working on our vision for mathematics instruction, our teachers at Adlai E. Stevenson were frustrated by the lack of authentic student engagement in our lessons. Our teachers felt that one of the primary barriers to student learning was our traditional acceptance of students sitting in rows. We had done an exhaustive review of the research and could not find any research that students sitting in rows would improve student achievement in any subject area—much less mathematics. One of our younger math teachers, Karen O'Staffe, asked, "Why do the students have to sit in rows?" and our small team of teachers investigating this issue agreed: "They don't." For our faculty, the mathematics instructional revolution finally began, and the results of our student mathematics performance began to soar.

We knew our expected instructional model for mathematics could no longer accommodate students sitting in rows. We had this deep-seated awareness that our students were just not engaged in the lesson due to our constant reliance on lecture or whole-group discourse.

As we were shaping our early work as part of the PLC process, we decided to pilot a different classroom desk arrangement structure using heterogeneously mixed groups of students who would be taught how to do work together in meaningful discussions about potential solution pathways to the work we were presenting. We called it *teams of four*.

The problem, of course, was that we had no idea how to manage the student teams and how to help them effectively engage. So we met together every Wednesday afternoon to share what was working, what was not working, and how to improve our expectations for the students to become a community of learners. Over time, our teachers—and our students—began to embrace a new culture for student learning.

Small-group discourse for teaching mathematics is a powerful instructional routine, and you will need to plan for it accordingly. When Jessica Kanold-McIntyre observes colleagues during a mathematics lesson, she notes that the most successful colleagues are able to *establish clear norms and expectations* for how students are to participate together during the small-group discourse and within a clearly allotted amount of time.

Remember that either extreme when planning mathematical discourse routines is not good. Too much time in small-group discourse can be just as damaging to student learning as too much time lecturing from the front of the room. The goal is to seek balance and to embrace the value of whole-group discourse as you need it.

Personal Story  **JESSICA KANOLD-MCINTYRE**

In my early years of teaching middle school mathematics and now with my experience as a principal observing mathematics instruction, small-group instructional routines are one of the most challenging aspects of lesson design. There were three types of barriers I faced during my early years of teaching that I also noticed in the mathematics instruction throughout our school.

First, small-group discourse with a mathematics task can cause feelings of losing control. To be successful at using small-group discourse, I became purposeful and intentional at planning to set up my expectations for how middle school students should work and share. These directed prompts involved both time and specificity in terms of my expectations for student communication during the mathematics task.

Second, the most accomplished mathematics teachers I have observed are able to use small-group discourse while preventing student conversations from veering off topic. These teachers have assessing and advancing prompts (see page 72 for more information) in place for orchestrating the small-group discourse as they monitor the room and provide engaging feedback prompts to the students.

Third, it is helpful to use clear transitions during the lesson between whole-group and small-group discourse activities. Various strategies, prompts, and structures help students persevere during small-group discourse and for the transitions back to whole-group discourse. These activities will be discussed in more detail on page 66.

From the front of the room, your opportunity to monitor and provide meaningful formative feedback to each of your students decreases. You are only one teacher for many students—sometimes thirty or more. When you utilize small-group discourse interwoven into your lessons, we recommend you use pairs of pairs—that is, twos and fours. For example, in a class of thirty-two students, you can use eight student groups of four (two pairs) for direct feedback from you and their peers and then for subsequent corrective action on the mathematical tasks presented during the lesson. Having eight teams of student feedback is more efficient than providing feedback to thirty-two students individually.

In *Building Thinking Classrooms in Mathematics*, Peter Liljedahl (2021) presents his finding that in grades K–2, student groups of two are optimal for small-group discourse. Younger students are still developmentally in a stage of parallel play and are still learning how to collaborate. Partnering is a great start to learning how to turn and talk, take turns listening, and agree and disagree respectfully. You and your team can choose games or higher-level-cognitive-demand tasks to explicitly teach collaboration and discourse skills. When designing the lessons, you and your team can build in time for practice and role play of the newly learned skills during active learning while the teacher is occupied with small groups. These small-group discourse activities can include designated spots during carpet time, partner talk, or roles during a game.

As a team, use figure 5.1 (page 62) to respond to questions about your current small-group or whole-group discourse routines.

Complete the teacher reflection (page 63) once you respond to the questions from figure 5.1.

Directions: Think about the nature of student discourse, generally, during your current mathematics lessons. As a team, discuss and share your ideas and strategies for student discourse during a lesson.

1. Consider the mathematics tasks you currently use to teach an essential learning standard. What percentage of the lesson time do you spend using whole-group discourse versus small-group discourse as students work through and learn from the solutions? Is the type of discourse your students experience balanced?

2. What type of structures do you like best for small-group student discussions? Teams of four? Partners? Threes? Explain why.

3. What is your biggest obstacle to including more opportunities for peer-to-peer small-group discourse time in your lessons? What can you do to remove that obstacle?

4. What is your best advice to ensure students persevere and stay engaged during whole-group discourse activities or direct instruction?

5. How do you ensure each student answers questions during a lesson and engages in learning through the tasks for both whole-group and small-group portions of the lesson?

Figure 5.1: Teacher team discussion tool—Student discourse during the lesson.

Visit ***go.SolutionTree.com/MathematicsatWork*** *for a free reproducible version of this figure.*

TEACHER *Reflection*

As you reflect on your answers to the questions in figure 5.1, list two or three lesson-design teaching practices you are now thinking about differently.

What actions in regard to student discourse, whole group or small group, might you be able to implement in your classroom tomorrow?

TEAM RECOMMENDATION

Balance Mathematical Discourse Routines

- In order to balance small-group and whole-group discourse, it's beneficial for students to be sitting in groups of four or some other team arrangement.
- In preK–2, during small-group discourse, it is important to teach students how to be good listeners, ask questions of each other, take turns, and learn from each other.
- Your best opportunity to see and hear how students are thinking, monitor that student thinking and learning, and provide formative feedback occurs when students are working in small groups on mathematical tasks as you monitor the student teams to see and hear what they are doing and thinking.

After you select the mathematical tasks and consider the best mathematical discourse routines to develop mathematical thinking, you also have to consider the structures and classroom practices that support mathematical discourse routines.

How to Implement Mathematical Discourse Routines

As you implement the mathematical tasks using either whole-group or small-group discourse routines, consider your current classroom culture and expectations for student engagement. Specific strategies for supporting student perseverance during whole-group and small-group time, and then managing student teams during small-group discourse, follow.

Supporting Student Perseverance During Whole-Group Discourse

As you move into the heart of the lesson using the mathematical tasks you have chosen for the students, you should ask yourself, "How will I check for student understanding during the lesson?"

Checking for understanding during whole-group discourse using verbal cues such as, "Did I go too fast?" "Isn't this an easy one?" "OK?" "Everyone see that?" "Who doesn't understand that?" or the often included rhetorical "Any questions?" is not sufficient. These particular cues set up two counterproductive conditions in your classroom: (1) they may not be accurate as a check for understanding and (2) they fall far short of your instructional goal to create and use a more formative feedback process with your students.

Reflect for a few moments on your questioning style during whole-group instruction.

Although it is difficult to do with much accuracy, there are two effective ways to check for understanding with feedback from the front of the room: (1) reflective summaries during presentation of new content and (2) effective questioning.

During a complex higher-level-cognitive-demand task, you can help students set up the initial investigation of the problem through your directed model, and then allow students some small-group time to discuss strategies for solving the task as you monitor their initial reactions and thoughts. This is a good time to circulate among students to find out if they understand the setup of the problem and possess the skills necessary to work a strategy of attack. While walking around, you receive and give feedback to pace the lesson more accurately. Students can work alone initially as they process their thinking, then compare and contrast solutions with peers on your command to do so. In the following personal story, Tim Kanold recalls his first steps to increase student-to-student discourse.

Personal Story  TIMOTHY KANOLD

I remember 1980 like it was yesterday. I was in my seventh season of teaching and my first year at West Chicago High School District 94. I attended our annual Illinois Council of Teachers of Mathematics (ICTM) meeting and received a thirty-page pamphlet from NCTM, our national organization. It was called *An Agenda for Action* (NCTM, 1980), and a primary recommendation was this: "*Teachers should provide ample opportunities for students to learn communication skills in mathematics. They should systematically guide students to read mathematics and talk about it with clarity*" (p. 12).

My initial response was, "I do not know how to do this. But I am for sure going to try," and my mathematics lesson process was forever changed. My first action was to eliminate student callouts to my questions during our whole-group discourse time in class, and engage my students in more peer-to-peer discussions during my questioning. I would redirect a student response for other students to evaluate, and they were surprised that I expected them to actually listen to one another and engage. I moved my body out of the front of the room and used more of a Socratic style as I helped students to rethink their mistakes. The result?

Students started to own their learning and not just depend on me for every moment of the lesson. In the beginning, they thought I was working them too hard because I was requiring them to actually *engage* in the lesson. Presenting solutions publicly (such as thirty students at a time standing at boards around my classroom) became a way of life during my daily lessons.

TEACHER *Reflection* 

In order to check for student understanding during the lesson, what types of whole-group questions do you find you ask the most?

Direct questioning, often done by you from the front of the classroom, is so important to effective instruction that it is one of eight research-affirmed instructional strategies that NCTM (2014) outlines in its influential publication *Principles to Actions*.

1. Establish mathematics goals to focus learning.
2. Implement tasks that promote reasoning and problem solving.
3. Use and connect mathematical representations.
4. Facilitate meaningful mathematical discourse.
5. Pose purposeful questions.
6. Build procedural fluency from conceptual understanding.
7. Support productive struggle in learning mathematics.
8. Elicit and use evidence of student thinking.

Questioning during the presentation of new content helps you assess student awareness, readiness, and understanding. An effective questioning cycle would expect all students to listen actively both to the question and to other students' responses, even though individual students are responding to specific questions. Can you use strategies that encourage *all* students to consider the questions you ask the whole class? Yes, but whole-group questioning can severely limit student engagement in class.

The questions you pose should require more than one-word answers. For example, the question, "What is the side length?" when the picture students are referencing shows five centimeters as the side doesn't promote meaningful or rich discourse. If you expect meaningful discourse, then questions must authentically engage students in discussion. For example, a question like, "What problems have we solved previously that are similar to this one? How are they the same? How are they different?" creates an open opportunity for students to explore and discuss.

A questioning cycle likely to result in this active engagement during whole-group discourse includes the following four steps.

1. Pose the question.
2. Provide wait time after each question to prevent student callouts (three or more seconds). Include time for a think-pair-share or write-think-pair-share to encourage all students to process the questions.
3. Select students randomly, making certain to include all students. Call on volunteers (raised hands) as well as nonvolunteers. Consider asking, "What did you and your partner (or team) discuss?" Who gets called on in class sends important messages about students' mathematical identities (Smith et al., 2017). "By carefully listening to and interpreting student thinking, teachers can position students' contributions as mathematically valuable and as contributing to the broader collective understanding of mathematical ideas" (Smith et al., 2017, p. 165).
4. Redirect the student response to other students for their judgment of correctness or for an extension of an answer. When you explicitly ask other students to comment on a student's work, you encourage a "diversity of views and strategies in the discussion" that, if effectively done, can "promote historically marginalized student populations or students who do not have a strong record of success in mathematics" (Smith et al., 2017, pp. 165–166).

During whole-group discourse, some students are reluctant to wait to be called on and like to call out an answer or response. Student callouts can be disruptive because these students control the pace and the direction of the classroom discussion, which then takes away from the learning for those students who need more time to process. If you notice during whole-group discourse that students are not immediately responding, you can also redirect the questions to small student groups for them to discuss before students are called on. This also promotes wait time for students to respond to other students' thinking.

When wait time goes unmonitored by you, your students usually receive less than one second to respond to a question. The use of wait time, at least three to seven seconds and up to ten seconds, accompanied by high-order questions results in students responding with more thoughtful answers and an increase in achievement (Cawelti, 1995).

A technique that also encourages wait time is to immediately follow a question with a phrase such as, "Raise your hand when you're reasonably sure of the next step or the response to my question." *Reasonably sure* indicates it is OK to take a risk and possibly be wrong. The following questioning sequence illustrates an effective questioning cycle. This particular cycle of questioning reduces student callouts as well.

Teacher: Class, what would be an example of a triangle with an area of 12 cm^2? Please draw and label a diagram on your paper and raise your hand to respond. [*The teacher walks away from the front of the room and monitors students as hands are raised.*]

Only John and Mia have such a triangle? Are there more? Three hands, four hands, anyone else? (The total wait time should be three to five seconds.)

OK, Alex, I notice you chose a right triangle. Please explain your diagram to the class. [*Alex does not have his hand up, but the teacher noticed his work and wants him to be able to share with the class.*]

Alex: I drew a right triangle with legs of length 8 cm and 3 cm.

Teacher: How many agree with Alex's example? Raise your hands! Who disagrees with Alex's example? [*This keeps all students accountable to listening to Alex's response.*]

Who can prove or disprove Alex's assertion . . . Karina? [*Karina raises her hand.*]

Karina: In a right triangle, the legs represent a base and a height. Thus, $A = \frac{1}{2}bh$ or $A = \frac{1}{2}(8)(3)$, which is 12cm^2.

Teacher: Do you agree with Karina's explanation? Raise your hand if you agree! Great! We all agree. As I walked around, I noticed all of you used a right triangle. Can you think of an example that is not a right triangle? [*The dialogue continues around this advancing question from the teacher.*]

This whole-group questioning cycle allows you to know which students are "with you" during the lesson. It also increases wait time (thus giving students more time to actually think about a response) by extending the conversation on a topic, which supports the learning for all students in class.

There are some students in class who may need up to fifteen seconds of wait time to respond. Thus, when you can use a question or multiple questions to *facilitate a peer-to-peer conversation* about the content rather than a correct answer followed by another question, you are able to allow more time for all students to make sense of the standard and the mathematics tasks you use in class.

Figure 5.2 provides a self-evaluation checklist on your whole-group questioning techniques. Individually, reflect on your current classroom discourse. Then, as a team, share areas of strength and areas where you can support each other as you continue to work on questioning.

As you look at your responses to figure 5.2, if your answer to every question isn't a 3 (always), there's room for improvement! Whole-group questioning and facilitation require thoughtful planning and purposeful teacher responses and guidance.

What types of questions are the best to use to help facilitate a whole-group discussion where more than one student response is heard? Are there strategies and sentence starters students can use to respond to each other?

In figure 5.3, you will find some sample teacher prompts and student sentence starters. As you read through the two lists, circle one or two new ideas you could bring back to your classroom to try.

Directions: Read each reflection question related to how you use questions in lessons. Identify whether you never, sometimes, or always employ the questioning strategy.

Reflection Question	1 Never	2 Sometimes	3 Always
1. Do I avoid asking, "Do you have any questions?"			
2. Do my questions promote total student involvement?			
3. Do I allow students to use think-pair-share before responding to questions?			
4. Do I require my students to act maturely when a student gives a wrong answer?			
5. Do I redirect certain questions and responses back to the entire class?			
6. Do I create a classroom environment that makes it safe for students to be wrong and even celebrate errors as a learning opportunity?			
7. Do I pause or give at least three to five seconds of wait time before calling on a student?			
8. Do I allow students to complete their answer before I jump in?			
9. Do I allow students to respond to another student's responses before I make a comment myself?			
10. Does my questioning give me meaningful input about the students' understanding of the concepts I'm teaching?			
11. When there are only a few hands raised to respond to a question, do I provide alternative ways to respond in order to get more students to participate?			
12. When only one student can answer a question, do I use this input and help others to understand and become involved in the question?			
13. Do I allow students to discuss ideas with their partners before asking a particular student to share ideas with the entire class?			

Once you complete the checklist, note both your strengths and weaknesses in your current whole-group question routines.
Compare your responses with a trusted team member and ask him or her if he or she would observe your lesson and let you know the level of student engagement during your whole-group questioning routines.

Figure 5.2: Teacher team discussion tool—Self-evaluation checklist on whole-group questioning.

Visit ***go.SolutionTree.com/MathematicsatWork*** *for a free reproducible version of this figure.*

Directions: Which of the following prompts do you currently use for whole-group discourse? List other ideas you use that are not listed.
Teacher questions to facilitate a discussion in whole-group discourse: • Do you agree or disagree with ___________ and why? • Who can add on to what ___________ said? • What is another way to say what ___________ said? • ___________, can you repeat what ___________ just said? • Can you say that in your own words? • Will this strategy always work?
Student sentence starters to support student mathematics talk in whole-group discourse: • I respectfully agree with ___________ because ___________. • I respectfully disagree with ___________ because ___________. • I understand what you're saying, but I have a different idea . . . • Another strategy you could use is . . . • Wait, I'm not sure I understand. Can you repeat what you just said? • So, I hear you saying . . . • How did you get that? • I noticed ___________; I wonder ___________.
List other questioning prompts or student sentence starters that you use:

Figure 5.3: Teacher team discussion tool—Whole-group discourse teacher prompts and student sentence starters.

Visit ***go.SolutionTree.com/MathematicsatWork*** *for a free reproducible version of this figure.*

For your feedback to be effective, even during whole-group discourse, your students must take action on the feedback they receive, either from you as the teacher or from their peers. This action is very difficult for you to elicit from the front of the room during whole-group discourse.

The next section brings into focus the only way to design your lesson so that you can see and hear *all* students in a more fully engaged classroom climate: small-group discourse.

Supporting Student Perseverance During Small-Group Discourse

You may be great at managing whole-group discourse. However, making it part of a *formative learning process* is almost impossible. You can do great checks for understanding, but the learning of a mathematics task needs to be formative for every student. It is best for students to experience the learning of mathematics as part of a small-group discourse activity in every lesson.

This is not a new concept. Since as far back as 1978, small-group discourse has been an expectation of expert mathematics instruction. According to Lev S. Vygotsky (1978), people learn complex knowledge and skills through social interaction.

Thus, accomplished teachers of mathematics have a built-in intuition with both deep and surface understanding of the fundamental purpose of a mathematics lesson: *student demonstrations of learning through productive perseverance, corrective feedback, and formative refinement of their work.*

Maintaining student perseverance during a mathematics lesson can be a very difficult task. There is quite a bit of evidence that *formative feedback with student action* during the lesson can help you overcome this daily student challenge in mathematics. Examining this evidence begins by understanding the nature of FAST formative feedback as part of the lesson-design process.

Team action 4, *analyze and use effective lesson-design elements to provide formative feedback and build student perseverance*, is all about moving learning forward. You and your team can provide deep support for student perseverance and proficiency with the use of effective formative assessment processes in your daily classroom instruction. Typically, you might not think of feedback and assessment as a support for learning *during* instruction, but effective assessment (particularly formative assessment processes as part of instruction) is critical to the learning process (Kanold & Larson, 2012; NCTM, 2014). According to Dylan Wiliam (2018):

> When formative assessment practices are integrated into the minute-to-minute and day-by-day classroom activities of teachers, substantial increases in student achievement—of the order of a 70 to 80 percent increase in the speed of learning—are possible . . . Moreover, these changes are not expensive to produce . . . The currently available evidence suggests that there is nothing else remotely affordable that is likely to have such a large effect. (pp. 160–161)

The formative feedback and assessment process is much more than just observing evidence of student learning (checking for understanding from the front of the room), which is at best diagnostic. For the process to also be formative, your students receive meaningful feedback from you or their peers (or both) during engagement with the mathematics tasks, *and* subsequently take action on that feedback. The formative assessment process is directly connected to the goal of student self-regulated learning and self-efficacy (Panadero, Andrade, & Brookhart, 2018). Tim Kanold shares a personal story on the next page about how one teacher designed a lesson for students to provide peer-to-peer feedback during the lesson by critiquing one another's work.

Take a moment to reflect on the role of formative feedback currently in your daily instruction.

TEACHER *Reflection*

Think about when you introduce a mathematics task during a lesson. How do you know when students get stuck, and what happens if they do get stuck?

Describe how you currently provide feedback to students while they are working on the mathematics problems you have selected for the lesson.

____

____

____

____

____

Personal Story  **TIMOTHY KANOLD**

I was working in a suburban Minneapolis school district for over seven months helping its teachers move out of rows and into teams of four for student discourse. At one point, I walked into a mathematics classroom of a teacher who had had a lot of success facilitating small-group discourse in student teams of four. Imagine my surprise to see her students sitting in seven rows of five deep!

I sat in the back, initially dismayed, but said nothing as I prepared to take notes on the student engagement levels during the lesson. The bell rang, and all of the students had a worksheet on their desks. On the teacher's command, they got started on the first math problem on the worksheet. I asked her for a copy and noted that it was a vertically stacked set of five questions, meaning that success in the five questions was somewhat dependent on understanding and success in each previous question.

But then the class got exciting! On her command, the students stood up, left their papers on the desks, and rotated back one desk and chair. The students in the back rotated to the front of their row. The students were allowed to fix the solution shown currently in question one, and then they moved on to question two. These rotations continued until each student returned to their original paper, where all five solutions awaited. They then had one last chance to reflect on the feedback they received and edit any parts of the five solutions.

Students turned in their papers and moved their desks back into teams. Students appreciated the opportunity to learn from one another and receive immediate feedback.

Essential Characteristics of Meaningful FAST Feedback

The Mathematics in a PLC at Work team uses the acronym FAST (*fair*, *accurate*, *specific*, and *timely*) to describe the essential characteristics of meaningful formative feedback explained by Douglas Reeves (2011, 2016) and John Hattie (2009, 2012; Hattie et al., 2017). Think of this as your attempt to provide FAST and corrective feedback to your students *during the lesson*. FAST feedback answers the following questions.

1. **Fair:** Does your feedback rest solely on the quality of the students' work and not on other student characteristics? This includes some form of student comparison to others in the class.
2. **Accurate:** Is the feedback during the in-class activity actually correct? Do students receive prompts, solution pathway suggestions, and discourse that are effective for understanding the mathematical task or activity as you tour and check for understanding?
3. **Specific:** Does the verbal feedback students receive contain enough specificity to help them persevere and stay engaged in the mathematical task or activity process? Does the feedback help them get "unstuck"—to advance their thinking as needed? (For example, "work harder on the problem" is not helpful feedback for a student.)
4. **Timely:** As you tour the room and listen in on peer-to-peer conversations, is your feedback immediate and corrective to keep students on track for finding a solution pathway?

These four characteristics describe how to give feedback; however, to have a magnified impact on student learning, you must also ensure students take action on the feedback. If during your best teacher-designed moments of classroom formative assessment processes, students fail to take action on evidence of their areas of continued difficulty, the learning cycle stops for the students.

During the lesson, take a close look at what your students are doing as the mathematics lesson progresses.

Are your students being expected to embrace their errors as they *reflect*, *refine*, and *act* on the in-class practice with their peers and with you?

According to W. James Popham (2011), the following occurs when teachers use formative assessment and feedback well during the lesson:

> It can essentially *double the speed of student learning* [emphasis added], producing large gains in students' achievement; and at the same time, it is sufficiently robust so different teachers can use it in diverse ways and still get great results with their students. (p. 36)

Similarly, John Hattie's research (Hattie et al., 2017) indicates that receiving formative feedback during lessons has a high impact on student learning (effect size of 0.75). The activity in figure 5.4 (page 71) is designed to facilitate your team discussion about the differences between checking for understanding as a diagnostic tool and using formative assessment processes as a student learning tool during mathematics instruction.

As you answer the questions and fill in your responses to the chart, consider how you can extend your various checks for understanding to become part of a more formative and FAST feedback process.

As teachers of mathematics, you and your colleagues should use various checks for understanding. However, realize this is only the starting point of the learning process for students during the lesson. Checking for student understanding at most has a diagnostic impact and helps you modify instruction as needed, but it has no real learning benefit for your students unless you extend the check for understanding to include a formative process with FAST feedback and action. This process, in turn, allows your students to embrace their mistakes (what will be their response during the mathematics lesson when they are not learning) and try again on that specific mathematics task during the lesson.

In-class formative assessment *processes* with student action on feedback support student perseverance on a task. As you and your team work together to implement more rigorous mathematics tasks that require students to use multiple strategies or representations and justify their thinking, student perseverance via meaningful and FAST feedback becomes essential to the learning process.

TEAM RECOMMENDATION

Use FAST Feedback During the Lesson

- Think about the type of feedback you provide students. Remember it must be fair, accurate, specific, and timely. Is there an area you need to improve on? What actions would make the feedback more FAST?
- Decide how you can shift your current use of checking for understanding to become more formative for students.

Promoting and using small-group discourse throughout your lessons encourages student perseverance while also maintaining rigor. It authentically promotes a community of learners who see each other as valuable resources and communicate about their ideas. It provides multiple opportunities for students to reason and make sense of the mathematics. However, to support discourse in your classroom, you must understand the demands students experience as sharers and listeners (Kazemi & Hintz, 2014).

Going beyond checking for understanding from the front of the classroom and moving into the more beneficial formative feedback process during small-group discourse is critical. Through robust student discourse, you and your team can engage students in meaningful and relevant mathematical tasks that align to the essential learning standard. This means supporting student-engaged explorations and discussions with peers. Use the teacher reflection to discuss your current use and facilitation of small-group discourse during your lessons.

TEACHER *Reflection* 

During small-group discourse, what types of instructional questions do you use to support student discussion and perseverance? What are strategies you use to support perseverance through the completion of a mathematical task?

______

______

______

Directions: As a team, use this tool to engage in a conversation about the differences between checking for understanding and formative assessment processes.

1. How do you currently check for student understanding on a daily basis? Write your responses in the left column of the following chart. For each strategy, indicate whether you use:
   - **W** for whole-group discourse (teacher at the front of the room)
   - **S** for small-group discourse (students working with their peers)
   - **B** for both
2. How do you and your colleagues describe the difference between checking for understanding and formative assessment processes in the classroom during instruction?
3. How can you implement formative assessment processes in the classroom each day? Remember for the process to be formative, students must take action on the FAST feedback they receive.

In the right column, explain ways that you could extend each check for understanding to become a more formative learning moment with action by students.

Checking for Understanding	Formative Assessment Process
Example: Using whiteboards or an electronic smartboard, students work in teams to find a solution pathway to the math problem presented and display solutions upon command from the teacher.	**Example:** Without judgment on the correctness of the solution, use the whiteboard responses to regroup students and ask them to each defend their solution, with a chance to correct potential errors and rewrite the solution based on feedback from peers.

Figure 5.4: Teacher team discussion tool—Checking for understanding as part of the in-class formative assessment process.

Visit ***go.SolutionTree.com/MathematicsatWork*** *for a free reproducible version of this figure.*

Questioning Effectively During Small-Group Discourse

To support the FAST formative feedback process (described on page 69) using small-group task exploration, you will need scaffolding (or "unstucking") prompts and advancing prompts ready to go for each task of the lesson. *Unstucking prompts*, also known as *assessing prompts*, are questions or statements you use to help students access the content or start the task if they are stuck. *Advancing prompts* are questions or statements you use to extend a task for a group of students who are demonstrating understanding on the current task. You will need to think of both the assessing and advancing prompts you will use with students before you implement the actual mathematics task in class. When teams are cocreating the assessing and advancing prompts, they are also planning for differentiation of mathematical tasks and structuring effective Tier 1 interventions, which will be explored more in chapter 7 (page 93).

How can you support student discourse during small-group instruction? First, you leave the front of the room and walk among the student teams. You listen in on their discussions, mistakes, and possible solution pathways. Second, you understand that your role during small-group discourse is to provide feedback. The crucial element of feedback is to improve your students' ability to perform the mathematical tasks (Wiliam, 2016). You are to facilitate the small-group classroom conversations and support student listening, engagement, and learning. You also learn quickly to provide feedback prompts that ensure all students have a chance to learn—not just the students who are first to talk in their group. Figure 5.5 (page 72) has sample preK–12 prompts you can use to support students while maintaining engagement. At the bottom of the chart, there is room for you to also note other prompts you might use.

Use the teacher reflection questions on page 73 to reflect on your use of assessing and advancing questions during small-group discourse.

<table>
<tr><th colspan="2">Assessing and Advancing Prompts to Ask Students During Small-Group Discourse</th></tr>
<tr><td>Prompts that help students work together to make sense of mathematics:
• Who agrees? Disagrees? Who will explain why or why not?
• Who has the same answer but a different way to explain it?
• Who has a different answer? What is your answer, and how did you get it?
• Can you please ask the rest of the class that question?
• Can you explain to your partner your understanding of what was just said?
• Can you convince us that your answer makes sense?</td><td>Prompts that help students learn to reason mathematically:
• Does that always work? Why or why not?
• Is that true for all cases? Explain.
• What is a counterexample for this solution?
• How could you prove that?
• What assumptions are you making?</td></tr>
<tr><td>Prompts that help students learn to conjecture, invent, and solve problems:
• What would happen if ________________? What if it did not happen?
• Do you see a pattern? Explain.
• What about the last one?
• How did you think about the problem?
• What decision do you think he or she should make?
• What is alike and what is different about your method of solution and his or hers? Why?</td><td>Prompts that help students connect mathematics, its ideas, and its applications:
• How does this relate to ________________?
• What ideas that we have learned before (prior knowledge) were useful in solving this problem?
• What problem have we solved that is similar to this one? How are they the same? How are they different?
• What uses of mathematics did you find in the newspaper last night?
• What example can you give me for __________?</td></tr>
<tr><td colspan="2">List other questioning prompts you use:

</td></tr>
</table>

Source: Adapted from NCTM, 2009.

Figure 5.5: Teacher team discussion tool—Assessing and advancing questions to ask students.

Visit ***go.SolutionTree.com/MathematicsatWork*** *for a free reproducible version of this figure.*

TEACHER *Reflection* 

Refer to figure 5.5. How could you use these types of questions within your own classroom?

What are one or two questions you could try using tomorrow?

How will you ensure all students take action on your feedback during their work together on a mathematics task?

When introducing and using a task during small-group work, it's important to set up clear expectations of how you want students to engage with each other and in the task. Then, as you walk around and support students, you can utilize the assessing and advancing prompts to help each student access the task.

When students have misconceptions during the lesson, they need to receive FAST feedback on the misconceptions, and then take action to correct their thinking or reasoning. Students then view reflection on and refinement of their in-class work as something they *do* in order to focus their energy and effort for future learning. As indicated so far in this section, a great time for this type of student reflection is during small-group discourse, as your students work together on various problems or tasks you provide, and you walk around the room, evaluate understanding based on what you see and hear, and then provide meaningful feedback to students and student teams. This process turns your checks for understanding into more meaningful formative assessment processes.

Thus, you and your students share the responsibility for successful implementation of in-class formative assessment practices. When your students can demonstrate *understanding through reasoning mathematically*, they connect to the essential learning standard for the unit and can reflect on their individual progress toward the learning target of that day's lesson. You support students' progress by using immediate and effective feedback during the daily classroom conversations, whether in whole-group or small-group moments, and then expecting your students to act on the feedback provided.

Monitoring Small-Group Actions and Results

You can use figure 5.6 (page 74) as an observational tool to collect information and provide feedback on your students' actions and results and their team engagement in the mathematical tasks and activities you plan for class. Ask a few fellow teachers to come in and help you with these observations as you need them, or there are additional ideas in chapter 8 (page 101).

As part of this formative process, you provide guidance and scaffold questions to support student learning and perseverance on the task. You also determine if you need to make adjustments in your whole-group instruction to develop students' reasoning skills. If a student team is ready, you encourage it to try the extension prompt or activity for the task as well.

You and your collaborative team should discuss how this tool or a modification of it might be useful in providing information about students' development of proficiency to reason abstractly and quantitatively during your mathematics lessons. As students are wrapping up their work on a task and it is time for you to select students or teams who will present the solution process to the class, follow these three steps Margaret Schwan Smith and Mary Kay Stein (2011) outline.

1. *Select* particular students to present their mathematical work during the whole-class discussion.
2. *Sequence* the student responses that you will display in a specific order.
3. *Connect* different students' responses and connect the responses to key mathematical ideas.

As you are selecting students to present (either randomly or based on their specific pathway), remind

	1—Working task but constantly stuck as a student team; does not generally take correct action on the feedback and prompts from the teacher	2—Working task through connecting to prior knowledge, engaging in conversations, and taking action to the scaffolding prompts from the teacher	3—Working task through engaging in accurate sense-making and reasoning, using multiple connections, and minimal feedback from the teacher	4—Working task is correctly done, and student team engages in an extension (advancing) prompt from the teacher during the small-group discourse	Next steps for the student team based on teacher observations
Student team 1					
Student team 2					
Student team 3					
Student team 4					
Student team 5					
Student team 6					
Overall observations of student reasoning and engagement:					

Figure 5.6: Team discussion tool—Walk-around formative assessment.

Visit ***go.SolutionTree.com/MathematicsatWork*** *for a free reproducible version of this figure.*

students to use *we* when explaining their team's solution (not *I*), as it helps to build the classroom as a community. Also think about how you can tell a story with the different strategies and solutions students produced during small-group discourse time. This is when the idea of sequencing student responses comes into play. The sequence can tie the mathematics together and allow you to connect the learning back to the essential learning standard for the lesson and unfold the mathematical story of the standard.

Note that in this "wrapping up the task" process, you do not go over the problem; rather, you allow student work and answers to be highlighted. You might assign a specific student as an expert for all similar problems of the same type or standard in the future.

Task implementation within small-group discourse, to be successful, will require you to manage certain classroom structures and routines every day. This is the focus of the following section of this chapter.

Building a Community of Learners

Creating a culture where students view their peers as resources is an outcome of small-group instruction. To effectively implement a balance of higher- and lower-level tasks, you effectively manage students during peer-to-peer discourse. When you set your student teams to work, do they know the expectation for teamwork, or do they rely on one or two students to get them started? Do they understand their rights and responsibilities?

Members of your student teams will need to build their norms for behavior, learn how to work effectively, and learn how to build consensus. Collaborative student actions (taught by you as their teacher) will ensure de facto accountability among students. Your students will need certain types of structures such as *rights, responsibilities, and norms* to support meaningful small-group mathematical discourse, engagement, and action.

Your collaborative team can create posters to share with students, or you can create these with your students. Figures 5.7 and 5.8 illustrate examples. Specifically, figure 5.7 offers rights and responsibilities for classroom discourse. Figure 5.8 offers norms of behavior and skills for small-group learning.

Before you can achieve access to the instructional learning targets for your lesson, it is necessary to ensure a safe classroom climate with clear expectations for student sharing and behavior. Students need to feel comfortable sharing their ideas and taking risks in front of their peers. This will also benefit multilingual students or students with identified learning disabilities. A space where learners feel comfortable first discussing ideas with their team members before sharing their thoughts more publicly with the entire class is a more inclusive environment. Take a few moments to reflect on your current norms and expectations for your students.

TEACHER *Reflection*

How do you currently create and then use class norms and expectations to help your students facilitate the sharing of ideas?

Rights	Responsibilities
Each student has the right to: • Ask for help • Be heard • Make a mistake • Express his or her thoughts on solution pathways • Learn the standard • Disagree with respect	Every student should: • Help others when asked by teacher or peers • Listen to the ideas of others • Take action on feedback • Be open to embracing errors • Seek consensus within his or her team • Work toward success with his or her peers

Figure 5.7: Sample classroom discourse rights and responsibilities.

Visit ***go.SolutionTree.com/MathematicsatWork*** *for a free reproducible version of this figure.*

Norms of Behavior	Skills for Small-Group Learning
Student team members should: • Listen carefully and with respect to one another • Contribute to the assigned team task • Ask other team members for help when needed • Help other team members who ask • Insist on logical persuasion before changing your mind	Student team members should: • Use quiet-conversation-level voices • Stay engaged and persevere through the mathematical task assigned • Ask peers for help, and then ask the teacher • Be supportive of each other • Ask for reasons or ask each other questions • Criticize ideas, but not the other students on your team • Have a sense of humor

Figure 5.8: Sample classroom norms.

Visit ***go.SolutionTree.com/MathematicsatWork*** *for a free reproducible version of this figure.*

One way to engage students in the process of developing norms for collaboration is to invite your students to respond to the following.

- Ask your students how they try to disagree with someone in a nice way.
- Discuss with your students what it means to make the conversations about mathematical ideas—and not about the person.
- Ask your students how team members should respond when someone on the team isn't participating.
- Ask student teams for strategies they can use as a team before they need to involve your support.

There are several cooperative learning structures or technology tools you and your colleagues can employ to create required participation, the key to engaging students in mathematical thinking—whatever the structure or activity (Johnson & Johnson, 1999; Johnson, Johnson, & Holubec, 2008; Kagan, 1994; Kagan & Kagan, 2009). Each of the following structures requires the use of a seating chart. See figure 5.9 for a sample seating chart template.

Consider the following four strategies.

1. **Use a structured seating chart:** This strategy makes it efficient to randomly call on a group to share or to call on specific students within a group. Since each student team has an assigned number, and each of the four students on the team is numbered from one to four, you can roll a die and call on student three from team five to present their team's solution. You can also use the structure to quickly organize the student work. For example, you can launch the mathematical task and state, "Student four in each team will lead the discussion when I give the signal."
2. **Use group and seat numbers to assign roles:** For example, all students who are number threes read the problem while number twos and fours lead the discussion, and number ones write down the solution being discussed.
3. **Randomly choose which student's paper you will collect:** This structure is one way to ensure all students keep on pace together and don't work ahead of other team members. A key factor to student team success is making

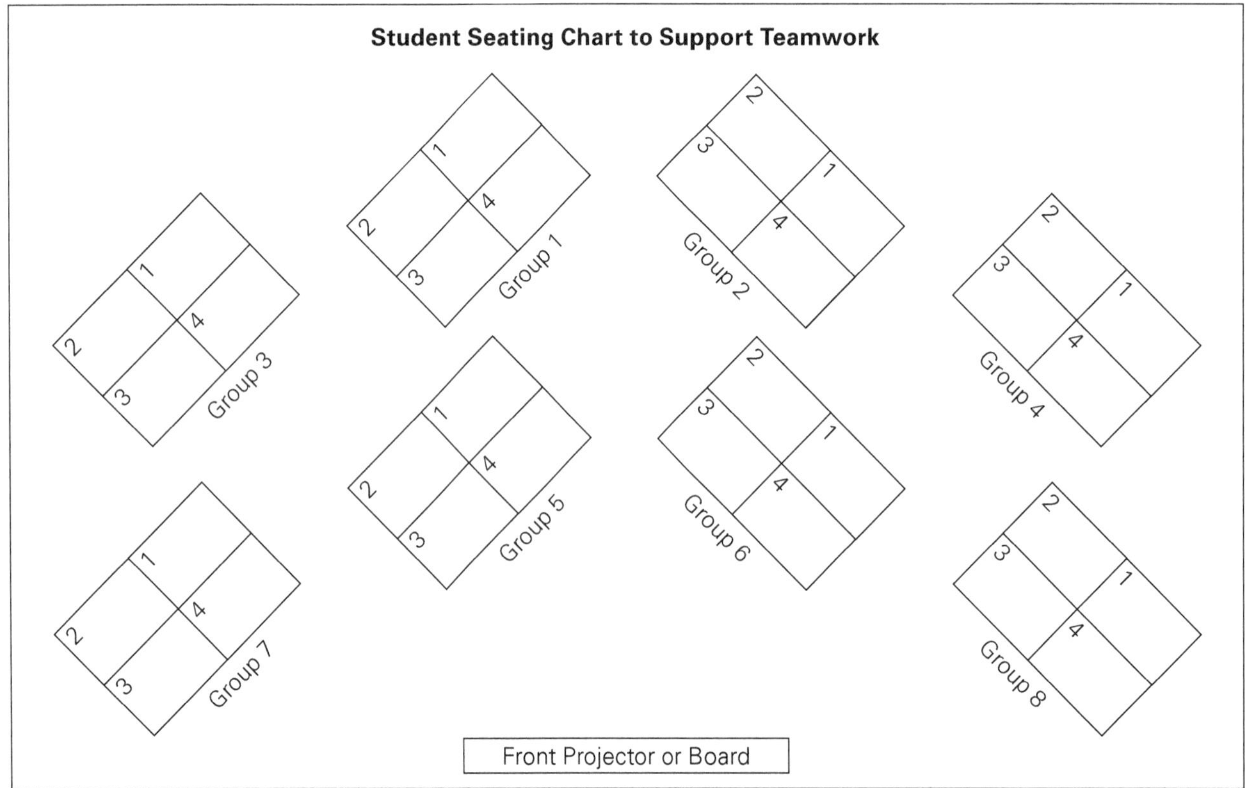

Figure 5.9: Seating chart to support the work of student teams during instruction.

sure everyone on the team corrects their misconception and understands a solution pathway to the mathematics task. If the students do not know which paper will be collected, they are more likely to ensure understanding for each student on the team. There are random number generators and group generators online that allow students to see the groups or students chosen are in fact random.

4. **Write notes on the seating chart:** This tool can help when recording scores or tracking student participation and collecting data for work completed and perseverance levels. It can also help to remind you of a student observation you made during the lesson.

The seating chart is a helpful tool for capturing data as well. For example, while observing your class, a colleague could use the following coding system to capture the types of questioning interactions within your class.

- Circle (○) = Student raises hand to ask question.
- Asterisk (*) = Student calls out answer.
- Square (□) = Teacher calls on student who doesn't have hand raised.

You could also use the seating chart to help track your movement around the room during small-group discourse. Use different colors depending on the type of small-group interaction you observe. For example, when you stop at a student group, use the following indications.

- Red = Supports the team by asking unstucking or scaffolding questions
- Green = Quietly observes student discussion and continues to the next group
- Blue = Supports the group by asking advancing questions for the mathematical task

Additionally, during small-group discourse, you and your team can utilize technology tools to encourage students to give feedback and learn from each other. You could have student teams complete a task on a shared document (such as a Google Doc, a Google Slide, a Jamboard, a Flip document, or an online Microsoft Word document) for three to five minutes, and then have them take five minutes to review the work of another student team and provide feedback. Take a few moments to reflect on how you manage your student teams during your lessons.

TEACHER *Reflection* 

How do you currently manage your student teams in your classroom? Do you use teams of two or more? What strategies and structures do you intentionally plan to help create a safe culture for sharing and risk taking by your students and their peers?

As a team, use the tool in figure 5.10 (page 78) to reflect on your current effort to manage your student teams.

As you and your team discuss your responses to the questions in figure 5.10, you may be unsure how to respond, especially if implementing student teams is new to you. That is OK.

You can then use figure 5.11 (page 79) to continue your teacher team discussion and reflect on next steps with your student teams. This figure contains responses to the questions in figure 5.10.

As you reflect on the strategies described in figure 5.11, you will likely realize it may be difficult to implement all the strategies immediately. However, it is important to start to move toward implementation of *student teams* while also implementing strategies and structures to create a learning environment where students take risks and share ideas during small-group discourse.

Directions: As a team, discuss and share your ideas and strategies for management of student teams during the lesson.

1. How do you currently organize your student teams? What size are the teams?

2. Think about your small-group or team activities, such as warm-ups, tasks, or closure activities. What type of management commands and prompts do you give student teams before they go to work?

3. What do you do while your student teams are working? What behaviors do you reinforce? How do you provide feedback if students are stuck? How do your teams know how to behave (norms)?

4. Once the student teams finish and you are ready to return to a whole-group focus, what verbal or nonverbal commands do you use?

5. Once there is whole-group focus, how do you summarize the problem or task with the whole group? Do you do it? Do you ask students to do it?

6. List any techniques you use to coach your students to listen to each other during whole-group discourse and small-group discourse.

7. Are there any other student team activities you believe support successful small-group discourse and learning?

Figure 5.10: Teacher team discussion tool—Management of student teams.

Visit ***go.SolutionTree.com/MathematicsatWork*** *for a free reproducible version of this figure.*

Directions: Possible responses to the questions from the management of student teams discussion tool (figure 5.10) follow. Highlight one or two responses in each category that you would like to start trying in your own classroom.

1. How do you currently organize your student teams? What size are the teams?

   It is imperative to not mix the students homogeneously during instruction because this can cause greater gaps in learning. Use only a heterogeneous mix for each team of students. I prefer to use groups of three or four to ensure students are included and heard during small-group instruction. I change the seating chart every unit so students get to work with all their peers. I also do random grouping throughout the unit to differentiate as needed.

2. Think about your small-group or team activities, such as warm-ups, tasks, or closure activities. What type of management commands and prompts do you give student teams before they go to work?

   - Nonverbal commands such as turning the lights out or raising your hand and expecting all students to raise their hand when they see you do it
   - Time commands such as, "You have five minutes; get started now."
   - Specific commands such as, "Work on the warm-up silently for one minute, and then share with your team."
   - Sharing commands such as, "Twos, you're the captains. After three minutes, ask each member of your group how to do the problem."
   - Timed directions such as, "In your groups, discuss for three minutes."
   - Desks out for silent work, and desks turned back in for working together
   - Directions such as, "Get started on the warm-up," with the invisible norm that students are to work alone and then pair-share
   - Visual directions—not just auditory ones—to support the different learners in your classroom

3. What do you do while your student teams are working? What behaviors do you reinforce? How do you provide feedback if students are stuck? How do your teams know how to behave (norms)?

   - Extensively tour the room and observe student work.
   - Redirect questions back to the whole group.
   - Tour student groups and listen to student conversations.
   - Tour and reinforce success, and encourage those who are not working with comments such as, "Right idea!"
   - Listen and kneel to notice work that is done well, and tell students to share with someone who is struggling with the problem.
   - Positively reinforce students who are helping other students, and encourage them.
   - Keep track of time and remind students with a verbal command such as, "You have two minutes left."

4. Once the student teams finish and you are ready to return to a whole-group focus, what verbal or nonverbal commands do you use?

   - Nonverbal commands such as walking back to the front of the room, placing a hand in the air for students to focus on, or turning the lights back on
   - Verbal commands that might include:
     a. "Pencils down."
     b. "OK, eyes back up front, pencils down, and listen to me."
     c. "Can I have your attention up front?"
     d. "Scholars ready!" "[Mascot name] ready!" or "Class, class!" with students replying "Yes, yes!"
     e. "I like what I am seeing from group ______ and group ______. They have their eyes up front, and they have stopped their conversation."
   - Whether using verbal or nonverbal commands, waiting until you have everyone's attention to talk to the whole group

Figure 5.11: Teacher team discussion tool—Management of student teams (with answers).

continued →

5. Once you have the whole-group focus, how do you summarize the problem or task with the whole group? Do you do it? Do you ask students to do it?
   - Tour the classroom to gauge everyone's understanding of the task you assigned so there is no need to go over the problem.
   - Have the students summarize; for example, all the threes go to the board to share (or share from their devices), a pair of students present the solution at the board, or a randomly selected student presents the problem.
   - Ask one student who has the correct answer or multiple students with different strategies and the correct answer to put their solution on the board as you continue touring the classroom and helping small groups with the problem.
   - Uncover the solution on the board and ask students to compare their small-group solutions with the one you present. What's similar about their strategy, and what's different?
   - Have groups write out their group solution on a piece of paper to present using a device in front of the class.
   - Have students show their work on whiteboards or chalkboards around the room so everyone can see all student work and the group can quickly examine and discuss any solution as it needs to.
6. List any techniques you use to coach your students to listen to each other during whole-group discourse and small-group discourse.
   - Explain to students that you expect them to listen to each other's responses.
   - Redirect questions back to other students in the class, and use a teacher explanation only when no other student can explain the solution or the reason for the solution.
   - Often restate the question or refuse to answer any student's question immediately, and try to allow other students to think about the question.
   - Rephrase the question if someone asks a question, and then ask the question of the whole class.
   - Make the question sound important by saying, "That's a great question. Who could answer this question or help _______ out?" Redirect the question to the class.
   - If an individual student asks a question, and it's important for the whole class to hear the question, state the question back to the whole class to think about a response. Possibly even have students discuss the question in their groups before answering in the whole group.
   - Eliminate student callouts by qualifying student questions with phrases such as, "Take thirty seconds to think quietly," "Place your thumb up on your chest when you have one solution," or "Raise your hand."
   - Record what teams say in terms of the mathematics and the social skills they demonstrate as they work on a task. Share these with the class as a celebration for great mathematics talk.
7. Are there any other student team activities you believe support successful small-group discourse and learning?
   - Use a student-evaluation form to gather information from teams about individual student participation to help support teams.
   - Plan activities in advance of the lesson to ensure they go well.
   - Do team challenges for friendly competition in class, which adds energy to the class—especially on days when students may seem sluggish.

*Visit **go.SolutionTree.com/MathematicsatWork** for a free reproducible version of this figure.*

Promoting productive small-group student discourse will allow your students to actively engage in the lesson, develop their mathematical voice, and increase their identity and agency. Students can self-assess how they are working together as you provide feedback to student teams on their ability to work cooperatively (Johnson & Johnson, 1999; Kagan, 1994; Kagan & Kagan, 2009). Figure 5.12 presents just such a student self-evaluation form you might use in class. It provides information on the collaborative process from the student perspective. You and your team can easily adapt this tool to an online form or survey where you ask students to reflect and respond.

Evaluator name: ____________________ Class: __________ Group number: __________			
Person 1	**Name:**		
This person is willing to help and explain when asked.	Never—0	Sometimes—1	Always—2
This person does their share of group work.	Never—0	Sometimes—1	Always—2
This person gives logical explanations as opposed to saying, "Just because" or "Believe me."	Never—0	Sometimes—1	Always—2
This person contributes to learning and makes it fun to work together.	Never—0	Sometimes—1	Always—2
Write the sum of the four scores for person 1 in the blank. (If person 1 is you, please circle the sum.) Sum = ________			
Person 2	**Name:**		
This person is willing to help and explain when asked.	Never—0	Sometimes—1	Always—2
This person does their share of group work.	Never—0	Sometimes—1	Always—2
This person gives logical explanations as opposed to saying, "Just because" or "Believe me."	Never—0	Sometimes—1	Always—2
This person contributes to learning and makes it fun to work together.	Never—0	Sometimes—1	Always—2
Write the sum of the four scores for person 2 in the blank. (If person 2 is you, please circle the sum.) Sum = ________			
Person 3	**Name:**		
This person is willing to help and explain when asked.	Never—0	Sometimes—1	Always—2
This person does their share of group work.	Never—0	Sometimes—1	Always—2
This person gives logical explanations as opposed to saying, "Just because" or "Believe me."	Never—0	Sometimes—1	Always—2
This person contributes to learning and makes it fun to work together.	Never—0	Sometimes—1	Always—2
Write the sum of the four scores for person 3 in the blank. (If person 3 is you, please circle the sum.) Sum = ________			

Figure 5.12: Student team processing and evaluation tool.

continued →

Person 4	**Name:**		
This person is willing to help and explain when asked.	☹ Never—0	😐 Sometimes—1	☺ Always—2
This person does their share of group work.	☹ Never—0	😐 Sometimes—1	☺ Always—2
This person gives logical explanations as opposed to saying, "Just because" or "Believe me."	☹ Never—0	😐 Sometimes—1	☺ Always—2
This person contributes to learning and makes it fun to work together.	☹ Never—0	😐 Sometimes—1	☺ Always—2
Write the sum of the four scores for person 4 in the blank. (If person 4 is you, please circle the sum.) Sum = ________			

Visit ***go.SolutionTree.com/MathematicsatWork*** *for a free reproducible version of this figure.*

Reflection on Practice

This chapter has focused on creating the structures for whole-group and small-group discourse and implementing effective mathematical discourse routines. The student management necessary to effectively transition from one type of discourse to the next will allow you and your students to more effectively respond to PLC critical questions 3 and 4: How will we respond in class when some students do not learn? and How will we extend the learning in class for students who are already proficient?

This is the exact reason to balance the use of small- and whole-group discourse. To respond to PLC critical questions 3 and 4, you must first know if students are learning during the lesson. That is impossible *if* you do not get away from the front of the room and you cannot hear what the students are thinking.

In this final story, Jessica Kanold-McIntyre relates an early experience as a middle school mathematics teacher.

Personal Story JESSICA KANOLD-MCINTYRE

I was doing my student teaching in the spring of 2004. I had some battles with my cooperating teacher about using teams of four to create an improved student-engaged climate for learning. She wanted the students to stay in rows as she did all of the work from the front of the room. We did battle.

It would have been great at the time to have the 2009 research by John Hattie, where he provides a great reminder of what matters most in teaching: *the role of the teacher is to constantly assess and have a pulse on where each and every student is in relation to the essential learning standards for the unit.*

So many times in my classroom I would jump in to summarize or "help" a student, which meant I would do all the talking instead of all the listening. I had to learn to let my students do the talking. I love that John Hattie's research validates my early understanding of good mathematics instruction.

TEAM RECOMMENDATION

Implement Mathematical Discourse

Consider the following recommendations with your team.

- Know that the cognitive demand and rigor of the tasks you choose matter in order to promote high-quality small-group discourse.
- Comprehend that small-group discourse allows the formative feedback process to come alive in your classroom.
- Consistently have your students in teams of two to four to establish a culture of collaboration and sharing. Take time to establish routines and expectations for how to hold a discussion and listen to each other.
- As students work and share ideas in groups, be sure to circulate and listen to student conversations so you can ask advancing or assessing questions and know how to sequence the whole-group summary of the task.
- Understand how the classroom culture and climate impact students' sense of safety for sharing, which in turn impacts students' learning and willingness to take risks.
- Establish classroom norms for whole-group and small-group behavior.
- Consider having students give you feedback on how comfortable they feel sharing and collaborating in your classroom. You could create a simple online survey for students to express their thoughts.

As you reflect on your instructional practices, eventually the time element to the lesson comes into play, especially as you use more small-group discourse as Jessica did. There is no reason students should be sitting in rows unless they are taking an assessment. You will not be able to cover quite as much mathematics content "ground" in each lesson. Yet class must end. There is one last element to consider within a classroom environment that fosters student perseverance and engagement. This is the way you bring the lesson home—lesson closure, as the final evidence of student learning that day. So now that the lesson is about to end, what do you do next? Surprisingly, in many preK–12 mathematics lessons, the answer is nothing. You are simply out of time. Just as you should warm students up at the start of the lesson with a prior-knowledge task or prompt, you should also help them cool down. Lesson closure is the focus of the next chapter, and the final lesson-design element.

Visit **go.SolutionTree.com/MathematicsatWork** for free reproducible versions of tools and protocols that appear in this book, as well as additional online only materials.

CHAPTER 6

Lesson-Closure Routines

Students learn when they are encouraged to become authors of their own ideas and when they are held accountable for reasoning about and understanding key ideas.

—*Randi A. Engle and Faith R. Conant*

How do you know if your lesson achieves its purpose? How do you know if your students can demonstrate evidence of learning the standard for the lesson? How do students know if they meet the learning target for the day? A lesson-closure routine allows students to summarize their learning from the lesson and you to determine the class's progress toward the essential learning standard.

Hattie (2009) references closure as follows:

> [as a way] to cue students to the fact that they have arrived at an important point in the lesson or the end of a lesson, to help organize student learning, to help form a coherent picture, to consolidate, eliminate confusion and frustration, and so on, and to reinforce the major points to be learned. Thus closure involves reviewing and clarifying the key points of a lesson, tying them together into a coherent whole, and ensuring they will be applied by the student by ensuring they have become part of the student's conceptual network. (p. 33)

Student-led summaries of learning progress during the lesson simultaneously yield accurate information on student progress and develop student self-efficacy and awareness.

Design of Lesson-Closure Routines

A lesson-closure routine is when you engage your students in a *student-led summary* of their learning at the end of the lesson. The benefit and purpose of the lesson-closure routines is to allow your students to take ownership of their own learning while also providing you with feedback about what they actually learned. This type of student-led closure is a review prompt of the content from the lesson or learning target for the day.

Student-led closure is different from a final check for understanding, such as asking students to do one more problem to wrap up the lesson. A student-led summary can occur before or after the exit slip, but it needs to reveal how the student *understands* and has processed the lesson activities.

The end of a lesson provides an opportunity for teachers to ask questions and plan for activities that tie together the content of the lesson. The focus of this time should be to help your students make connections and help them address misconceptions you have observed during the lesson. It can also reveal the students' feelings about the lesson and their level of confidence for understanding the essential learning standard.

As a team, share different ways that each of you plan for and enact lesson closure. This is a great time to learn from each other and share ideas. Use the following team discussion tool (figure 6.1, page 86) to support these professional conversations.

As you choose the way you will close the lesson, be sure to connect the discussion prompts you provide back to the learning target or targets for the day, and the essential learning standard for the unit. The closing question, prompt, or activity you choose should also support an appropriate way for students to have time to reflect on the overall lesson experiences so they can accurately summarize and connect their learning.

In the personal story on page 86, Tim Kanold shares an example of a student-led summary.

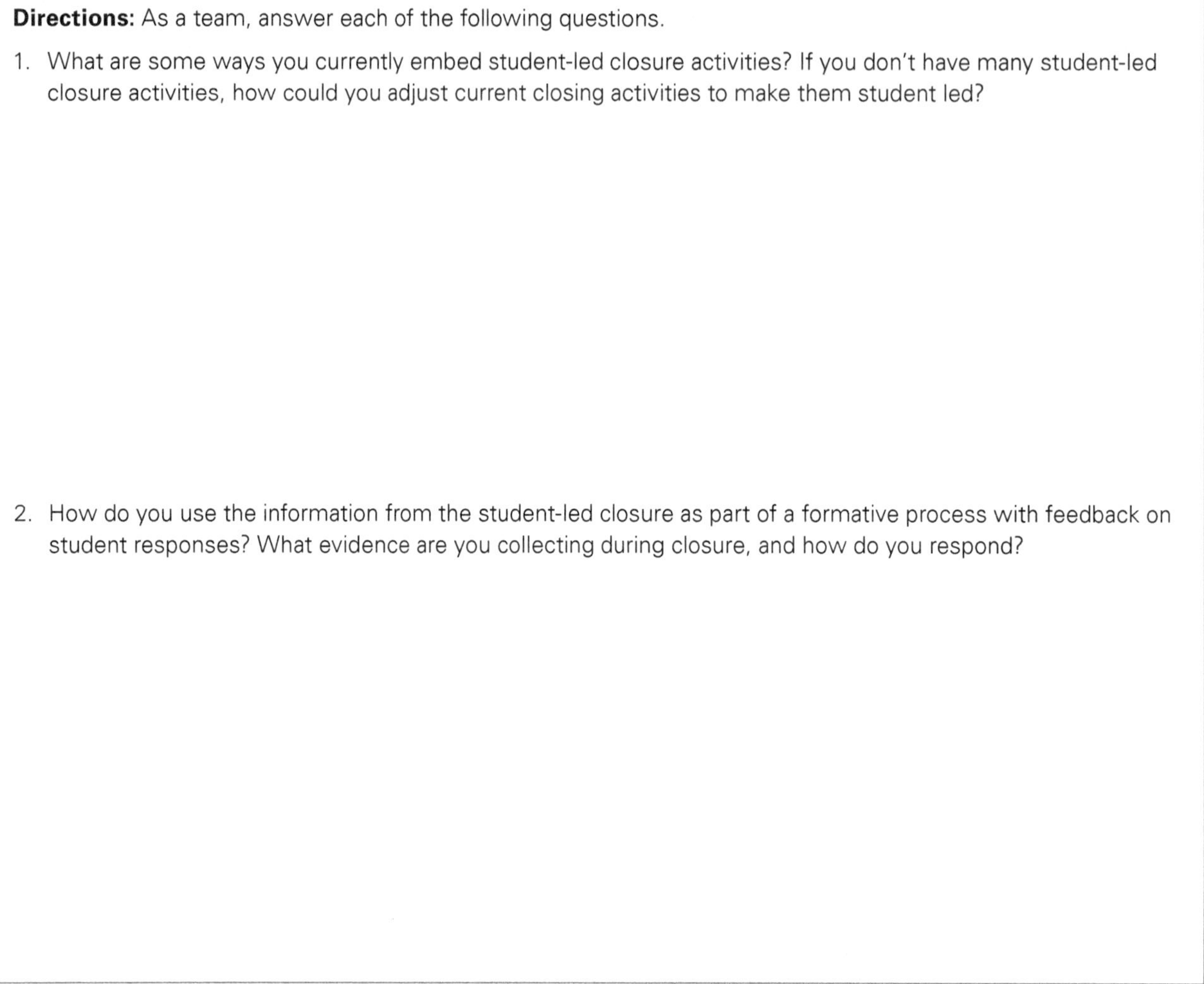

Directions: As a team, answer each of the following questions.

1. What are some ways you currently embed student-led closure activities? If you don't have many student-led closure activities, how could you adjust current closing activities to make them student led?

2. How do you use the information from the student-led closure as part of a formative process with feedback on student responses? What evidence are you collecting during closure, and how do you respond?

Figure 6.1: Teacher team discussion tool—Lesson-closure reflection.

Visit ***go.SolutionTree.com/MathematicsatWork*** *for a free reproducible version of this figure.*

Personal Story TIMOTHY KANOLD

At Stevenson, we had a mathematics teacher, Mary Layco, who would eventually become recognized by the Illinois State Board of Education as an outstanding mathematics teacher in Illinois. The first time I watched her teach a math lesson, she passed out paper plates to every student and had them stand in circles with five students per circle. This was with about seven minutes left to go in the lesson. She asked the students to write down in the middle of the plate their answer to this prompt: "What is one thing you would like to know more about from today's math lesson?"

She then asked them to rotate the plates one person to the left and provide an answer to the question on the plate. After five rotations (about one minute each), each plate was back to the original owner with five answers to the question! I remember thinking, "Wow! What a great idea!" and immediately used the strategy with my students. She did not know it at the time, but Mary was creating intentional student-led summaries as part of her lessons.

Here are some sample closure prompts or questions.

- "How did mathematics tasks ________ help you understand the learning target ________ today?"
- "What is one thing you understand about [*learning target*] that you did not know this morning?"
- "How would you explain [*learning target*] to your friend if they were absent today?"
- "What is one thing you would like to know more about based on today's lesson?"
- "On a piece of paper or in your journal, summarize the lesson in three sentences."
- "List in your notebook the one thing that made the most sense to you today."
- "When someone asks you what you learned in math today, what will you tell them? Start with 'I can . . .'"
- "What part of the math lesson was most confusing for you today? State it in one sentence."
- "Based on this unit so far, name the math standard you would like to practice more before the unit ends."
- "Go to your online Google Form and work in pairs to describe what you learned today and what confidence you have in doing the mathematics."

Remember, mathematics lesson closure routines are quick reviews to connect your students with what it was they learned (the daily learning target) and allow you to gather information on their understanding and progress as you plan the next lesson.

Furthermore, closure promotes clarity to your mathematics lesson, as you bookend your opening (Why are students learning to do these math problems today?) with your closing (Does the student have clarity on what they actually learned?). This, in turn, promotes the idea of student agency as part of the daily owning of students' learning, and their understanding of what they were to have learned.

Reflect for a few moments on closure prompts you use as part of your current lessons in the teacher reflection.

TEACHER *Reflection* 

How do you currently plan for student-led lesson-closure activities as part of your lessons each day?

What is your favorite closure activity? List different strategies and ideas.

The academic work of analyzing and summarizing what has been learned as part of the lesson each day should be done by your students—not by you. Closure allows your students to summarize main ideas, potentially answer questions posed at the beginning of the lesson, and demonstrate their own meaning making. How did the lesson turn out from their point of view—not your point of view? Next up, you will explore how to implement student-led lesson-closure routines.

TEAM RECOMMENDATION

Design Lesson-Closure Routines

- Student-led summaries of learning allow you to gain information on student progress and confidence at the end of a lesson.
- Student-led summaries should be connected to the daily learning targets and essential learning standard for the unit.
- Be sure to allow an appropriate way for students to reflect regarding their overall lesson experience as they make connections to their learning experience for that day.

How to Implement Student-Led Lesson-Closure Routines

Being *intentional* in your planning to include a lesson-closure routine at the end of every lesson—the final lesson-design element—is another way to embed formative assessment into your daily lesson plans. If students' current conceptions are incorrect or they have misunderstandings, a student-led summary gives you great feedback on next steps in future lesson decisions. How do you currently facilitate a student-led summary? Should it be a whole-group, small-group, or individual activity? What does it look like in the classroom? What would you see or hear students doing during this time? Mona Toncheff provides some insight into the importance of this ending aspect of your lesson in her story.

Personal Story  MONA TONCHEFF

I will never forget my first coaching experience when I was a new teacher. The instructional specialist came into my classroom to observe me as a first-year teacher. She provided positive feedback on student engagement and then asked me one simple question, "How do the students know that they met the learning target for the day?" At that point in time, I didn't have a response.

As I reflected on the lesson, I was thinking about the evidence of student engagement as students were working efficiently with their peers to make sense of the mathematics. The students looked like and sounded like they met the target; however, I had not given the students an opportunity to reflect on their own learning. I assumed students would individually take responsibility to reflect on the learning for the day. What I didn't realize as a new teacher is that not all students enter into my classroom as reflective learners.

During my first five years in the classroom, I had to work diligently to give students time at the end of each lesson to wrap up their learning for the day so that when they left my classroom, they could articulate what they spent their class time learning. When I became more fluent in closure strategies, my students also grew into reflective learners, asked better questions of their peers, and understood how to use the learning targets as a pathway for their learning.

As you and your team members think about how you want to implement lesson-closure routines, remember there are many ways you can have students lead the closure activities. The key is that students are *doing* the reflection and summary around the learning target for the day, and that it is not teacher led. Lesson-closure routines are an informal way to take the pulse of student understanding and provide you with critical information about how effective your lesson was at supporting student learning of the daily learning target that aligns to the essential learning standard of the unit. In his story on page 89, Tim Kanold describes his early experiences on the issue of closure with his mathematics articulation committee.

Personal Story TIMOTHY KANOLD

At Adlai E. Stevenson, four to five times a year, we would gather our Mathematics Articulation (MAC) Task Force. As the director of mathematics, I had the responsibility of chairing these professional development events as we tried to bring a K–12 coherence to our mathematics efforts. After several years of meeting together (schools often sent different teacher representatives each year), we realized our K–12 mathematics lessons were "incomplete stories."

We had become pretty good at starting the story of the lesson for each day. By then we had made a K–12 commitment to students' learning mathematics in teams of four, even buying new horseshoe-style desk furniture with separate student chairs. We were also good at opening the story with warm-up activities and connections to the learning standard (back then we called them *SLOs—student learning objectives*) for the day.

To some extent, the middle of our mathematics lesson story was pretty good too. We were becoming much better as teachers of mathematics at using meaningful mathematics tasks and engaging the students in discussions. We did not have a good ending to the story. Ever. As a body of K–12 teachers, we just thought if the students seemed to be doing OK, then they knew what they were doing. There was no need to end the lesson's story.

To prepare for my meeting with the MAC, I spent two months traveling from school to school and observing teachers of mathematics. I would ask students one question at the end of the lesson (in elementary school) or end of the class period (in middle school and high school): "In thirty seconds or less, tell me what you learned today." Not surprisingly, even in the lessons that went extremely well, the students' responses were widely different, sometimes inaccurate, or inarticulate. They just did not know the purpose and story of the mathematics lesson.

As teachers, we were at fault. We were not providing any opportunity for the story to end and for the students to own the ending as they reflected on their work for the day. I gathered several teachers from different grade levels, and the five of us (I was teaching a class of calculus AB) started experimenting in our classes on various endings. We intuitively knew that how students felt (level of confidence) was related to how they could summarize what they knew from that day's lesson. We asked students to keep journals and write their responses to our prompts, and then share with a partner. These were simple student reflection questions, not a complicated set of exercises.

When our MAC Task Force met, we submitted our findings, each of us relating our experiences and our surprise that our students did not remember or view understanding of the mathematics lesson the same as we did. We began our journey of making sure the story of the mathematics lesson would always have an ending they would write.

Figure 6.2 (page 90) presents possible lesson-closure routines that allow you to address student learning from the lesson while also keeping the focus on a *student-led* summary of the lesson rather than a teacher summary. These types of routines provide great data about where students are in their progression of understanding the essential learning standard. They provide solid evidence for you and your team in your planning process.

Routine	Description	How the Routine Is Formative
Student reflection	Use a specific question that ties to the content from the class. The question is of higher cognitive demand to assess true understanding at a conceptual level rather than just a procedural level.	For the student reflection to be formative, there must be teacher and student action on the information. For example, the teacher must review answers, sort the results into groups (got it, almost got it, not yet), and then give each group a specific problem to begin the lesson the next day.
Student team summary	Have groups write or draw what they learned for the day and share with the class.	For this activity, students who are listening to the summary should ask the group questions and provide feedback—like a Socratic discussion. The feedback to students is immediate, and the teacher can document group understanding to use in planning for the next day.
Questioning—small group or whole group	Ask students questions such as: • "Why?" • "Could you explain it another way?" • "How does this connect with ____________?"	The questions must be crafted to facilitate a conversation that provides feedback on student understanding. Through the questioning process, the feedback is immediate to students, which helps them shape their understanding in the moment. As a teacher, you are responding formatively by listening and choosing your next question based on the answers from the students.
Gallery walk	Capture the complex task students worked on in class on poster paper, and hang the paper around the room. Students walk around the room giving feedback such as: • "I wonder . . ." • "I like . . ."	After the gallery walk, the class provides feedback and students make adjustments to their work based on the feedback (that day in class or the next day).
Student presentations	Have student groups present their work from a task from class or present their summary of the lesson.	During the presentation, students record specific content each group mentions and offer a note about one thing they like and one question they have. The groups get this feedback to review and adjust their thinking.
Voting with feet	Pose agree-or-disagree questions, and ask students to move to one side of the room or the other depending on whether they agree or disagree.	What would normally be a check for understanding can turn into a fun classroom debate between differing sides by having students explain why they chose their answer. Students then revote after the debate.
Nonverbal check	Using a scale of 1–5, ask students to hold up the appropriate number of fingers to their chest to indicate their comfort level and confidence with the learning target for the lesson. (Alternatively, you can use thumbs-up or thumbs-down.)	Prepare multiple questions and have them ready to go as a check-in with students. After the nonverbal check, regroup students for a re-engagement activity based on self-reported responses. In those new groups, provide students differentiated instructional tasks with specific feedback as needed.
Voting tools (like Google Forms, Schoology, Edmodo, Haiku Learning, Formative, and so on)	Give online quizzes where students get their results immediately and you can see all student results.	This is a great way to capture actual data for each and every student in an efficient way, but it can be difficult to make feedback formative. Some tools allow the teacher to type a response directly back to the student. The data can also be used to regroup students for a differentiated warm-up activity the next day.
Online discussion forums (like Schoology, Google Classroom, Edmodo, Socrative, TodaysMeet, and so on)	Have students participate in online classroom discussions where they share their thinking, read classmate explanations, and learn from each other.	This strategy is a great way to use technology to provide students with a forum to communicate about their mathematics learning outside the classroom. Provide specific questions tied to essential learning standards at the end of a class, or use the forum as a way for students to ask each other questions about homework, and so on. This requires clear expectations for student behavior and some monitoring by the teacher, but it can provide positive support.

Figure 6.2: Teacher team discussion tool—Sample lesson-closure activities.

Visit ***go.SolutionTree.com/MathematicsatWork*** *for a free reproducible version of this figure.*

Reflection on Practice

This chapter has focused on creating lesson-closure routines for student reflection at the end of the lesson using *student-led summary* prompts. This routine allows students to consolidate their learning and make connections to the mathematical story of the lesson or unit. This lesson-design element is the final step in your design process for high-quality lessons.

TEAM RECOMMENDATION

Implement Lesson-Closure Routines

- As a team, commit to embedding student-led closure into your daily lessons.
- Take your current activities and strategies used for closure and examine how you could make them more student centered with formative feedback to the students.
- As a team, consider how you can use student-led closure to gather evidence of your students' demonstration of proficiency on the daily learning target. How can you and your team use this evidence to prepare a lesson for the next day or unit?

Chapters 1–6 have included identifying the essential learning standard for the unit and the related daily learning target for the lesson, and identifying the prior-knowledge and mathematical language routines necessary for student success. They have also focused on the importance of carefully choosing the mathematics tasks that you believe will support student learning of the standard, and balancing these tasks cognitively. You have learned about the need to choose various tasks for your lesson and the nature of the communication and discourse for each task, carefully balancing whole-group and small-group discourse and creating student-led summaries to close every lesson.

The last part of the Mathematics in a PLC at Work lesson-design tool (see the appendix, page 112) and the way you reflect on the evidence of effectiveness of the lesson set up the final two chapters of this book. You will examine how you should respond to student progress during the lessons of a unit and use that input in your planning decisions for the next lesson, the next unit, or sometimes the next year. You closely examine Tier 1 interventions—the interventions you provide as part of instruction—as you ask, "What do I need to do differently to help every student learn?"

In the next chapter, you learn how to respond to student progress during the lesson with high-quality Tier 1 mathematics intervention.

Visit **go.SolutionTree.com/MathematicsatWork** for free reproducible versions of tools and protocols that appear in this book, as well as additional online only materials.

CHAPTER 7

High-Quality Tier 1 Mathematics Intervention

In the end, all learners need your energy, your heart, and your mind. They have that in common because they are young humans. How they need you, however, differs. Unless we understand and respond to those differences, we fail many learners.

—*Carol Ann Tomlinson*

Students learn mathematics at different rates, yet every student is expected to demonstrate proficiency for the essential learning standards in your grade level or course. Therefore, you and your teacher team plan units, create common assessments, and design lessons with every student in mind and discuss how to engage all students using those learning tools. Douglas Fisher, Nancy Frey, and Carol Rothenberg (2011) suggest that "intervention is an element of good teaching" (p. 2).

Intervention begins in the classroom with the effective instruction you plan every day. As noted previously, one of the challenges to high-quality instruction is that different students need different supports at different times in any given lesson (Hattie et al., 2017).

To ensure all students learn at grade level or course level and above, your teacher team and school will create systems and structures to provide interventions (and extensions). A three-tier framework for intervention is often called a *multitiered system of supports* (*MTSS*) or *response to intervention* (*RTI*). Austin Buffum, Mike Mattos, and Janet Malone (2018) and Richard DuFour and colleagues (2016) explain that Tier 1 intervention should be provided for all students *during* core instruction.

Tier 2 intervention is for small, targeted groups of students and happens during an additional time in the school day, and Tier 3 is provided for individual students needing even more time and support to accelerate learning to grade level and above. For more information on the three tiers of RTI, visit AllThingsPLC.info to read a blog post by RTI expert Mike Mattos (2016; www.allthingsplc.info/blog/view/335/connecting-plcs-and-rti).

For the purposes of this book related to your teacher team's design of high-quality mathematics lessons, the focus is on Tier 1 interventions. For more information about your team's design and use of Tier 2 interventions, see *Mathematics Assessment and Intervention in a PLC at Work, Second Edition* (Schuhl, Kanold, Toncheff, et al., 2024). The ultimate goal is for you and your colleagues to design lessons that provide opportunities for immediate interventions (during the lesson) in Tier 1 with an eye toward reducing the number of students needing Tier 2 or Tier 3 supports.

In *Learning by Doing*, DuFour and colleagues (2016) describe Tier 1 intervention actions that occur during the school's core instructional program as you provide students access to the essential grade-level learning standards *and* adjust to more effective instructional strategies for your students. Much of Tier 1 intervention involves your and your colleagues' intentional planning of daily lessons using the criteria in the Mathematics at Work instructional framework while embracing a "look in the mirror" mindset as you ask, "What can *I* do differently during instruction to grow student learning?"

Think about the core of your instructional decision-making and planning process. How does your lesson design support and align to the team's unit plans (see the *Mathematics Unit Planning in a PLC at Work* series)? Remember, you are the first line of defense for struggling mathematics students, and your during-the-lesson decisions either provide or do not provide an opportunity to accelerate learning at grade or course level for students who are not there *yet*.

A well-planned lesson does not focus solely on what is being taught; it also includes opportunities to ensure students are learning during the lesson. Pérsida Himmele and William Himmele (2021) share the following:

> Teachers who excel at what they do constantly monitor learning and recalibrate their teaching according to what they see. . . . It's simply what it means to teach well. These teachers don't wait for exam results to tell them who needs help mastering content, because they continuously work at staying aware of who does and doesn't understand. They gauge the extent of students' understanding, and they take steps to provide support, redirection, and reteaching when it is needed. (p. 57)

Interventions during core mathematics instruction include your differentiated responses to student learning. You provide a differentiated response by using scaffolding prompts *during the lesson* to help students think of other pathways to enter into or ultimately solve a higher-level-cognitive-demand task (see figures 4.3–4.7, pages 52–54). Ongoing in-class interventions are at the heart of your Tier 1 instruction and proactively address improved student learning, which in turn prevents your students from needing Tier 2 or Tier 3 interventions and support.

As you and your colleagues intentionally use the six lesson-design elements that appear in chapters 1–6 to plan and carry out your mathematics instruction and flexibly adjust to student learning during your lessons, more students will demonstrate initial success on your unit assessments.

Tier 1 Intervention Routines

The purpose of Tier 1 intervention is to provide *all* students access to essential grade-level and course-based content and ensure effective daily teaching that results in student learning. Effective Tier 1 intervention happens during prior-knowledge routines, through use of lower- and higher-level-cognitive-demand tasks, in whole-group and small-group settings, and throughout all parts of any lesson. When planning each part of the lesson, ask, "What is the teacher doing?" and "What are the students doing?" Consider how to provide time and utilize tasks that allow for you to check for understanding and orchestrate formative assessment processes (see chapter 4, page 45, and chapter 5, page 57) throughout the lesson. Powerful core instruction described by Gayle Gregory, Martha Kaufeldt, and Mike Mattos (2016) includes five components.

1. Clearly articulated essential standards
2. Success criteria for mastery (proficiency)
3. Evidence-based instructional practices
4. Meaningful, relevant, and student-centered instruction
5. Twenty-first century skills

You and your colleagues can use formative assessment information—data; information on students' prior knowledge; information about students' diverse language and cultural backgrounds (beliefs, perspectives, histories, and attitudes); and so on—to address students' learning needs, especially as you give them feedback on their progress and they take action with their peers during the lesson.

Throughout the lesson, you can observe student reasoning by listening to student-to-student discourse or viewing student work on whiteboards, on poster paper, or in notebooks, to name a few options. Cassandra Erkens, Tom Schimmer, and Nicole Dimich (2018) share that your observations of student work provide an effective and efficient way to gather feedback and inform your next steps during instruction. Observing what students are doing and saying gives you insight into your students' thinking and requires you to interpret students' actions and words in ways that lead to their growth.

Students' learning of mathematics during a lesson increases when they are given opportunities to understand the grade-level or course-based concepts (see chapter 1, page 11) through connections to prior knowledge (see chapter 2, page 23). How can a lesson address conceptual understanding? Often students make sense of and deepen their understanding of mathematics concepts through higher-level-cognitive-demand tasks. For Tier 1 intervention, you consider how to differentiate entry and exit points to the higher-level-cognitive-demand tasks by utilizing formative assessment strategies during instruction. Procedural fluency tasks often follow, then, through practice with application. You are not making the standard less rigorous; rather, you are making learning more accessible to more students by the time the mathematics lesson ends.

Bill Barnes explains how his teacher teams began to work together to rethink this issue in his school district.

Personal Story BILL BARNES

When we began working with our mathematics teams to build an understanding of lower- and higher-level mathematical tasks, we found that a common teacher misconception emerged. At least one teacher on each team, often more than one, would say, "If students cannot even solve these lower-level problems, how can you expect us to ask them to solve a harder problem?" The misconception centered on a belief that mathematical thinking had to be developed strictly from procedure to application . . . but never the other way around.

Our team initiated monthly professional learning that featured a fishbowl learning laboratory so that teachers could observe students' responses and engagement when introduced to higher-level-cognitive-demand tasks prior to learning procedures. Over time, teachers began seeing the value in engaging students in a worthwhile mathematical task as a strategy for student engagement in deeper mathematical thinking.

Tier 1 intervention can also occur during small-group instruction using intentional stations or centers for a portion of any lesson. After observing student learning in class or through lesson exit tickets, you might create targeted stations (by learning target with appropriate tasks) for small-group rotations. Each station becomes a small-group minilesson experience as you address common misconceptions and provide feedback to students.

As discussed throughout this book, clear classroom expectations, intentional groupings of students by standard, and student learning through discourse are all important aspects of creating a safe environment for a culture conducive to improved student learning. Consider how you and your team will use in-class Tier 1 intervention. Use table 7.1 and the teacher reflection to consider the responses to learning you can have and the potential impact on students.

Table 7.1: Teacher Team and Student Characteristics of Effective Tier 1 Core Instruction

Teachers and Teams	Students
Create a safe space for learning. (See chapter 5, Building a Community of Learners, page 74.)	Students feel comfortable making mistakes and learning from each other.
Create opportunities within tasks to build social skills. (See chapter 4, page 45.)	Students learn how to listen to each other, ask questions, encourage each other, and constructively formalize their thinking.
Develop positive student-teacher relationships.	Students are engaged and involved in the learning process.
Collaboratively determine what each and every student will know and be able to do at the end of every unit. (See chapter 1, page 11.)	Students have a clear understanding of the learning pathway for the unit and lessons.
Choose mathematical tasks that are open-ended, low-threshold, high-ceiling tasks to provide multiple access points. (See chapter 4, page 45.)	Students can begin the task using the skills that they have access to, using a less complex approach to a more complex way to solve the problem.
Create a student-centered community of collaborative learners. (See chapter 5, Building a Community of Learners, page 74.)	Students view their peers as resources to support their learning.
Use multiple representations, technology, and manipulatives to make sense of the mathematics.	Students see the importance of multiple representations and how to make connections and create flexible mathematical thinking.
Provide opportunities for students to take ownership of their learning. (See chapter 6, page 85.)	Students feel empowered to own their learning and understand they can take action on the feedback provided to improve their understanding.

TEACHER *Reflection*

Of the teacher team responses to student learning listed in table 7.1, which actions are strengths from your current practice? Which actions are not as consistent? Choose two on which to focus during lesson design. List some specific actions you could take to ensure you implement the in-class Tier 1 intervention strategies this year.

How to Implement Tier 1 Intervention Routines

Tier 1 intervention is about your instructional response to student learning during class as you teach the mathematics standards using different strategies in order to impact a greater number of students during initial instruction. Tier 1 is also about your future lesson adjustments based on data revealed once your students showcase their learning on common mid- or end-of-unit assessments (see *Mathematics Assessment and Intervention in a PLC at Work, Second Edition* [Schuhl, Kanold, Toncheff, et al., 2024]). The student work data may reveal instructional practices that were more effective across your team and should be used in the future. The data and student work might also reveal possible common misconceptions, errors, or mistakes to address with specific students in a Tier 2 setting (see *Mathematics Assessment and Intervention in a PLC at Work, Second Edition*) or in a Tier 1 setting if the content is prior knowledge students will use in future lessons.

Of course, the expectation is that before students engage in your common unit assessments, they have received FAST feedback (*fair*, *accurate*, *specific*, and *timely* feedback—see page 69) from you during daily lessons. It is also expected that you and your colleagues have worked collaboratively to benefit from the data you have collected during the in-class formative feedback process as you make final instructional adjustments to support student preparation for the common unit assessment.

Figure 7.1 shows an example of an observational chart members of your teacher team can use to gather evidence of student learning after an initial assessment. Your team discussion leads to instructional strategies to use in future lessons so students learn the essential learning standards in the unit before the common end-of-unit assessment.

Student Name	Misconception	Error	Mistake	Valid Solution Strategy With an Accurate Answer

Figure 7.1: Sample form for collecting data observations.

Visit ***go.SolutionTree.com/MathematicsatWork*** *for a free reproducible version of this figure.*

As you review your student assessment performance data together, you and members of your team record the type of mistake, error, or misconception and check the student use of a valid solution strategy with an accurate answer. Your team then shares and discusses the data results to determine any learning standard misconceptions that need to be addressed in future lessons, throughout the unit, and into the next unit.

Tier 1 intervention is provided during your use of prior-knowledge warm-up routines and during the lesson instructional task routines based on how you and your team plan the assessing and advancing questions (chapter 5, page 57) for each mathematical task or activity. When you create a differentiated response to student learning during core instruction, the design of the tasks and the supporting questions for exploration lead to teacher conversations on how to differentiate during instruction. Carol Ann Tomlinson and Cindy A. Strickland (2005) share that there are three ways to differentiate during instruction.

1. The *methods* that students use to access content
2. The *process* that students use to practice learning the essential standards
3. The *products* that students create to demonstrate proficiency of the essential standards

When teams are designing lessons and selecting the mathematical tasks aligned to the essential standards, teams can also intentionally plan how to differentiate during the lessons based on the evidence of student learning. Figure 7.2 includes examples of strategies for all three ways of differentiating.

Methods	Process	Products
• Minilessons (video) with different worked-out examples • Highlighted or underlined text in word problems • Different problem-solving strategies • Different manipulatives, models, or representations for the solution pathway	• Opportunity to work alone, in pairs, or in small groups • Group roles when in small groups • Choice of tasks to complete • Choice of entry points into the mathematics task • Various types of graphic organizers	• Product options that respond to varied interests or learning profiles • Varied timelines or check-in points • Varied criteria for success (for example, from novice to professional) • Varied audiences (in age, background knowledge, size, and so on) • Some choice of questions on common independent practice

Source: Adapted from Strickland, 2007.

Figure 7.2: Examples of differentiation strategies.

As you read through figure 7.2, what specific strategies do you and your team members currently use to provide access toward learning the standard or standards for the lesson? Are there additional instructional routines that you could employ during core instruction to provide more access?

Jackie Acree Walsh (2022) discusses the importance of questioning when developing formative assessment strategies in a lesson, specifically stating the following:

> Questioning that results in dialogic feedback does not occur by accident; rather, it results from intentional and strategic teacher design of formative lessons and learning environments. Formative lesson design, best accomplished in collaboration with colleagues, is integrated into daily lesson planning and includes two primary focus areas: (1) questions to activate student thinking and responding and (2) structures intended to optimize the number of students who communicate their responses. (p. 105)

The use of open-ended questions (figure 5.5, page 72) and the utilization of pair-share and small-group discussion within whole-group discussions (see page 72) provide opportunities to gather evidence of student learning and make adjustments in the lesson if needed or allow students to adjust their understanding.

Tier 1 opportunities during the lesson occur when students make their thinking visible for you to quickly assess their learning and devise a plan. Students may write on whiteboards or paper so other students (or you) can give FAST feedback to grow their learning as they think through the mathematical task.

To effectively teach, intervene, and extend learning during core instruction (Tier 1), the instructional block of time must be protected (Buffum et al., 2018). While protecting instructional time is a school issue, it may be something your team needs to advocate for to ensure student learning of mathematics. Additionally, protecting instructional time means engaging students in learning from the start of the class time to the very end.

Sometimes in lessons, you and your teacher team may forget to leave time for students to process their learning, to pause and think and make connections. Bryan Goodwin and colleagues (2023) share:

> When your students say that something does not make sense or they don't get how to do something, what they're really saying is that their brains need more time to pause and process their learning—often with their peers. Or they may be saying that they need some coaching—opportunities to be observed and receive feedback while trying new skills. (p. 83)

The small-group discourse discussed in chapter 5 (page 57) allows you to observe students in order to determine if more processing time is needed. You might ask students to summarize what they have learned so far, or you might develop a prompt for them to reflect on the essential learning standard for the lesson.

Another strategy to make in-class Tier 1 interventions effective is to include students in their learning. Based on feedback throughout the lesson from you and other students, how well have students learned each essential learning standard? Look again at the student-led closure strategies, and consider how students can reflect on their learning progress by using a reflection tool or tracker throughout a unit. Help your students to know which essential learning standards they have learned and which they have not learned *yet*. (See *Mathematics Assessment and Intervention in a PLC at Work, Second Edition* [Schuhl, Kanold, Toncheff, et al., 2024].)

Reflection on Practice

Jeffrey Benson (2021) discusses explicit and implicit formative assessment strategies. *Implicit* formative assessments are learning experiences with feedback through observations and student discussions. *Explicit* formative assessments are when students do the work given in a planned task during the parts of the lesson when students are expected to show their thinking. Use the teacher reflection to connect the results of student learning to your daily instructional routines and in-class Tier 1 interventions. Consider the implicit and explicit formative assessment strategies you use to respond to student learning in real time.

TEACHER *Reflection* 

How can you continue to meet the needs of each and every student this year? What intentional in-class instructional responses to student learning can you make as a team to intervene and support students who are struggling within a unit?

When answering PLC at Work critical question 3—How will we respond when some students do not learn?—you and your teacher team respond in Tier 1 and Tier 2 learning experiences. Tier 1 is about instructional moves and routines you can make in class during lessons to impact student learning. Tier 2 is a more robust and targeted re-engagement experience for students needing additional time and support to learn an essential learning standard, often implemented across your team at an alternate time of the school day outside of the mathematics block of time or class period.

Mathematics Assessment and Intervention in a PLC at Work, Second Edition (Schuhl, Kanold, Toncheff, et al., 2024) provides deeper guidance on your Tier 2 intervention response, when in-class Tier 1 interventions are not sufficient for student learning of each standard.

TEAM RECOMMENDATION

Implement Effective Tier 1 Interventions

- As a team, be fierce about access to rigorous and engaging learning opportunities for students.
- Discuss how you will each monitor and respond to evidence or a lack of evidence of student learning as part of your in-class instructional strategies.

It is your team's responsibility to continually evaluate the effectiveness of your core instruction and adapt instruction as needed to meet the needs of each and every student. It is your collective response to evidence of student learning that will make the most significant impact on student success. The final chapter explores how collaborative teams can engage in cycles of continuous improvement to improve the effectiveness of instruction.

Visit **go.SolutionTree.com/MathematicsatWork** for free reproducible versions of tools and protocols that appear in this book, as well as additional online only materials.

CHAPTER 8

Analysis of the Effectiveness of Mathematics Instruction

Improving something as complex and culturally embedded as teaching requires the efforts of all the players, including students, parents, and politicians. But teachers must be the primary driving force behind change. They are best positioned to understand the problems that students face and to generate possible solutions.

—*James W. Stigler and James Hiebert*

According to Harvard researchers, "What predicts performance is what students are actually doing" (City, Elmore, Fiarman, & Teitel, 2009, p. 30). Analyzing student learning requires more than just looking at the end result of the lesson or unit of instruction. Collaborative teams also analyze how students are interacting during lessons. The in-depth analysis of students in action reveals more than what teachers predict students do or do not understand when they are analyzing student work (City et al., 2009).

Deborah Loewenberg Ball, Mark Hoover Thames, and Geoffrey Phelps (2008) share that teaching is a professional practice that requires knowledge and skill beyond what appears in the essential learning standards for mathematics. Not only do teachers need content knowledge (the most essential learning standards) and the progression of the learning targets in a given unit, but team members also need pedagogical knowledge on how students make meaning of the content and what instructional strategies most impact student learning for student engagement, as well as the scope of student needs (Ball et al., 2008).

You can intentionally create opportunities and structures to make your instructional decisions more transparent for your collaborative team by using peer observation. As a teacher, you are usually so focused on in-the-moment decisions that it is nearly impossible to simultaneously and carefully assess the impact of your decisions on in-lesson student actions and interactions. Because peers can focus on in-lesson student actions and interactions, teacher peer observations can help your team grow your instructional routines and practices. Currently, it is rare to observe a school culture where teachers of mathematics consistently observe one another teach the lessons they design together. Yet in highly effective school cultures, this type of learning from and support of one another becomes the norm.

In this chapter, you will explore team reflection strategies that support building the shared knowledge you and your teacher team need in order to design effective daily lessons for a positive impact on student learning.

Team Reflection on Evidence of Student Thinking

At the end of the lesson, how do you and your colleagues know whether your instructional practices are effective? How do you know which to replicate in a future unit, as an intervention strategy, or during the same unit next year? Teacher teams can use either student assessment results or evidence of student thinking during instruction to evaluate the effectiveness of the Tier 1 instruction.

On page 102, Jennifer Deinhart shares her personal experience of how her team used team reflection to improve professional practice.

Personal Story  JENNIFER DEINHART

Teaching fifth graders the relationship between fractions and decimals is a concept that influences the learning during that school year and beyond. Early in the unit, it was common for us to start exploring equivalent fractions and decimals by making fair shares with friendly fractions and base-ten blocks.

As a team, we agreed on the learning targets and the common assessments. However, we had not discussed the depth of the instructional routines for the task. A teammate of mine took the exploration a bit further and implemented the task by asking her students to work with a variety of contexts and tools. Each small group was asked to represent their answer in both fraction and decimal forms, and the task was posed with the following questions and tools.

CONTEXT FOR THE FAIR SHARE	TOOL USED BY THE STUDENTS
If 8 people had 1 whole dollar to share equally, how much would everyone get?	Dollar bills and coins
If 8 people had 1 whole candy necklace to share equally, how much would everyone get?	Beaded number line to 100
If 8 people had 1 whole piece of bread to share equally, how much would everyone get?	Base-ten blocks

Since small groups were using different tools to represent the equivalent values, the students were able to make comparisons and connections with the tools. It led to a deep understanding of what happened with the leftover hundredths and how that portion needed to be represented by thousandths. During both group and whole-class discussions, students were coming to conclusions that they could justify rather than simply shading in an amount on graph paper that showed $\frac{1}{8}$ of 100.

When the team reviewed the exit ticket results, we asked her why her students had a stronger understanding of the concepts. She shared her more in-depth approach, and due to the positive results on the exit ticket assessment and the retention over time, our entire team moved to use the task and tools in this way. I have since coached multiple fourth- through sixth-grade teams about the power of connecting different contexts and tools to make sense of a big idea.

Analyzing Assessment Evidence for Tier 1 Intervention

Your use of common assessments during a unit and at the end of a unit informs not only which students need additional time and support to learn the essential standards, but also which instructional practices worked most effectively across the team.

Through your analysis of student performance evidence at the end of a mathematics lesson against each essential standard, you can consider trends in student learning, including common misconceptions, in order to better design your responses for the next lesson.

When analyzing data from a common assessment, you and your colleagues can take three steps as you reflect on the data and then use those reflections to take action on possible in-class interventions to your instructional strategies to teach the mathematics standard.

1. Select an essential learning standard, and then use student work and the corresponding feedback to identify what sets proficient and exceeding students apart related to their work.
2. Next, look at the work of students who are close to proficient, and begin to see if there is a common misconception or a misunderstanding to target as a catalyst for possibly moving students to proficient since you cannot always reteach the entire standard.
3. Finally, look at trends in student work from those students who are far from proficient to identify what to target to improve learning.

In the personal story on page 103, Timothy Kanold shares his experience as a member of a mathematics teacher team that leveraged assessment results to adjust and improve instruction.

Personal Story  TIMOTHY KANOLD

As we slowly began to embrace the idea of using student performance from our assessments to impact the nature of our lesson design, I established an expectation that every mathematics teacher team in our district would examine their data together and then share instructional strategies that seemed to be working best for their students as a way to adjust our instruction (the mathematical tasks used, strategies for teaching the standard, and our use of formative in-class differentiation for learning).

I was also a teacher on our AP calculus AB team. As it came time to teach the unit on volumes by cross-section standard, we examined both our local data from the assessment the year before and our national data on the AP exam (it is a major topic on the essay portion of the exam). Our analysis of student performance on the standard revealed this standard was an extremely poor performer, for a few others and me. But one teacher had reasonably good results on our local assessment and on the national test.

We were aware that a Tier 1 response would not be to go on teaching it the way we had done in the past. Even though we felt we were good teachers, our *strategy* for teaching the standard was not very effective. We knew our professional Tier 1 response would be to try a different strategy. So, Chris (our team leader) showed us her use of a modeling project to build the solids, and then reveal the formulas. We all used her activity, maybe a bit adjusted, but nonetheless this same visual representation, and the intervention worked. It was so effective we never had a problem with student proficiency on this standard again.

You can use the questions in figure 8.1 to evaluate your current progress toward connecting your in-class instruction and adjustments (the formative assessment process work) to the mid- and end-of-unit assessment expectations from one unit to the next or one year to the next.

Directions: Select your next common assessment. Reflect on each statement individually, and then discuss as a team several responses taking place during instruction that will prepare students to be successful on that common assessment instrument.

	Strongly Agree				**Strongly Disagree**
	5	**4**	**3**	**2**	**1**
1. Teachers on our team use a variety of formative assessment strategies to encourage students to use and persevere through both standard and invented strategies.					
2. Our students represent their learning in multiple ways, such as with number lines, tables, graphs, functions, equations, and other types of models.					
3. Teachers on our team demonstrate that the answer to a mathematics problem or task is not all that matters in learning; reasoning and explaining why and how are also expected.					
4. Our students improve proficiency with both concepts and procedures through experiences with higher- and lower-level-cognitive-demand tasks.					
5. Our students increase their problem-solving and reasoning skills by solving problems together; hence teachers present students with higher-level-cognitive-demand tasks without lowering the demand during instruction and require students to justify their responses and reasoning for the solution pathways to such tasks.					

Discussion notes:

Figure 8.1: Teacher team discussion tool—Connecting in-class Tier 1 intervention to the end-of-unit assessment.

*Visit **go.SolutionTree.com/MathematicsatWork** for a free reproducible version of this figure.*

Analyzing Evidence of Student Thinking During Instruction

In addition to reviewing assessment evidence to evaluate the effectiveness of Tier 1 instruction, you can also open up classrooms to encourage teachers and team members to observe student thinking in action. Thus, learning becomes more visible to you, your colleagues, and your students. In this section, you will explore two protocols that support building the shared knowledge teacher teams need in order to design effective daily lessons: (1) instructional rounds and (2) lesson study.

Instructional Rounds

Instructional rounds, also called learning walks, are a process for teams to engage in collective inquiry around their Tier 1 instruction. Typically, teacher observations in the classroom are either informal or formal evaluations by an administrator. Through the evaluation process, the teacher being evaluated is typically the only one expected to learn. With instructional rounds, the expectation is that everyone participating is learning from the experience (City, 2011).

The purpose of instructional rounds is similar to the purpose of medical rounds that occur during residency. During the instructional rounds, a team of teachers walks from classroom to classroom collecting data on a focused instructional task (for example, questioning, specific instructional strategies, engagement, classroom management, and so on).

Once the rounds are complete, the team of teachers debriefs its observations, provides just-in-time and specific feedback to its peers about the levels of student engagement in the lesson, and formulates action steps based on the evidence of student learning.

Instructional rounds or learning walks can be grade level based, course based, or cross-curricular, depending on the instructional focus. Instructional rounds are beneficial because they (City et al., 2009):

- Break down the isolation of traditional teaching and make teaching a public event
- Require collaborative teams to focus on a common challenge that each team member wants to improve
- Build capacity with instructional practices and develop a common language related to instruction
- Provide a way to hold teachers accountable for improving instructional practices with feedback in a nonevaluative manner

Instructional rounds promote reflection about the instructional strategies used and their relationship to student learning actions. The primary purpose and strength of the instructional round is the reflection that occurs during the debrief following the observations, the discussions that stem from the observations, and the actions teachers take as they identify next steps (Marzano, 2009). However, without teacher action on the feedback, the instructional round focus will have limited impact on student learning.

As you begin the instructional round process, you will need to establish norms for the instructional rounds. Since the purpose of instructional rounds is to observe student learning, norms ensure observation and feedback are tied to evidence of student engagement and learning. The following are some basic norms for instructional rounds.

- When observing in a mathematics classroom, refrain from conversations with the other observers.
- Refrain from assisting students or interrupting the lesson.
- Collect factual observations about student thinking, reasoning, and actions that are nonevaluative.

The final point is very important: when you and your team are collecting evidence, your comments and observations during the debrief need to be factual rather than judgmental. For example, you may observe a group of students working on a task and see one of the students is on his phone underneath the desk. A judgmental statement would be, "One student was completely off task and not following the school rules." A more factual statement is something like, "I observed three student teams. One team had one student not engaged in the task." Notice the difference in the tone?

Before you begin using instructional rounds or learning walks, you and your team also need to identify an instructional focus goal. What part of your mathematics instructional vision would you like to explore? What is your school improvement goal? What is the instructional focus from your SMART goal? For example, one of the research-informed instructional strategies from *Principles to Actions* (NCTM, 2014) is to pose purposeful questions. If you were to observe a mathematics lesson with your team, would each member be in agreement about what posing purposeful questions looks like or sounds like in a classroom? Start by describing both the teacher and student actions that your team members may observe and what you would like to see during the observations.

By defining what you and your team are looking for in Tier 1 instruction, you and your team members can co-create observational rubrics that are aligned to your vision of instruction. When you and your team collect evidence of student learning during the instructional rounds, you and your team will be empowered to define personal and team actions to create instructional routines that best align with your vision and equally support student learning (Wilson, 2022).

For additional information and resources on instructional rounds, see *Mathematics Coaching and Collaboration in a PLC at Work* (Kanold, Toncheff, et al., 2018). You can also visit **go.SolutionTree.com/MathematicsatWork** or scan the QR code in this book to access the free reproducibles for instructional rounds.

Lesson Study

Lesson study is a particularly powerful collaborative tool. Research shows that lesson study is very effective as a collaborative protocol with a high impact on teacher professional learning (Gersten, Taylor, Keys, Rolfhus, & Newman-Gonchar, 2014; Stigler & Hiebert, 1999). Your team can also use this process to participate in collective inquiry during the unit.

Lesson study is a professional learning model that exemplifies the importance of the collaborative process. Lesson studies are opportunities for teams to collaboratively:

- Co-plan the "best mathematics lesson ever" using ideas from all the teachers on the team
- Observe the mathematics lesson in action together
- Revise the mathematics lesson based on feedback from the team and evidence of student learning
- Teach the mathematics lesson again to see if the revisions had the predicted impacts

Lesson studies can take many forms, but the premise of a lesson study is not just to focus on improving an individual teacher; rather, it is to collectively improve methods of teaching across a team of teachers. In 1999, James W. Stigler and James Hiebert, in *The Teaching Gap: Best Ideas From the World's Teachers for Improving Education in the Classroom*, identified the seven critical steps of a lesson study as follows.

1. **Define the problem:** What do we want to learn about by engaging in the collaborative process?
2. **Plan the lesson:** What resources will we need to focus on our instructional goal?
3. **Teach the lesson:** Which teacher will teach the lesson the first time?
4. **Evaluate the lesson:** What evidence do we observe and collect during the lesson?
5. **Revise the lesson:** What might we change in the lesson to improve student understanding?
6. **Teach the revised lesson:** Who will teach the lesson the second time?
7. **Evaluate and reflect again:** What might we still change in the lesson to improve student learning?

Lesson study differs from instructional rounds because during lesson study, you and your team are observing a team-created lesson; during instructional rounds, the lesson can differ, but the instructional focus is common.

For additional information and resources on lesson studies, see *Mathematics Coaching and Collaboration in a PLC at Work* (Kanold, Toncheff, et al., 2018). You can also visit **go.SolutionTree.com/MathematicsatWork** or scan the QR code in this book to access the free reproducibles for lesson study.

As teams are codesigning the lessons, teachers are able to learn from each other during these critical planning conversations. Sarah Schuhl shares her experience helping a teacher team codesign a lesson using the Mathematics in a PLC at Work lesson-design tool in the appendix (page 111).

Personal Story  SARAH SCHUHL

To help a second-grade team more fully understand shared teacher learning, we planned a lesson together for two-step word problems, an important concept for students to learn and one that is often difficult to teach. The teachers identified the standard, essential learning standard, and daily target to frame the lesson. In the discussion, one teacher realized that the emphasis is not on two steps, but rather that students reason and think through the task to its completion. Some students might think about the problem in two steps, and others might complete a step mentally and show one complete step. The planning idea was that students would start and complete the task in different ways and then be able to share their solution strategies and answers.

Armed with this information, we talked about what students had learned prior to the lesson that would give them the tools and resources to work through the two-step tasks. Students had solved one-step word problems with the unknown in all parts of an equation and added and subtracted within 100.

The team decided to launch the lesson with a prior-knowledge activity using a missing addend word problem that required double-digit addition or subtraction to solve. Students would work together in groups to share ideas for solving the task. Next, we identified the academic language students and teachers needed to intentionally use in the lesson with sentence frames or on a word wall for clear discourse in class.

The teachers then identified three tasks from their curriculum to use for the lesson—one that would be modeled using their problem-solving template. (What do you know? What do you need to know? Draw a picture to explain the problem. Show your thinking to solve the problem. Write your answer.) They would then ask students to work in pairs to solve the next task and set a timer to keep students focused.

After two minutes, they would have students mix up and share their starts with one another for feedback before going back to their original partner and working or modifying their solution. After another five minutes, pairs would share their thinking in their original teams and then share selected solutions with the class.

The third task would be completed by groups of four and posted around the classroom. Teachers carefully thought about how students would use discourse and visible thinking to give feedback to one another while the teacher also could be available to work with students and groups for formative feedback.

One teacher realized that she had only been modeling word problems and then having students work on them, not teaching students to see each other as resources and learning partners. Throughout the discussion, we also talked about how to differentiate the learning by being prepared with a manipulative for students who might struggle, and extending the learning by asking students to *create* a word problem requiring a student to add and subtract that would have an answer between thirty and thirty-four, thus creating a question that teachers could later use in class.

Through planning a lesson together, each teacher had a new takeaway related to strengthening their own lessons on a daily basis. The team also gained a deeper understanding of the standard students needed to learn, which allowed teachers to give better feedback to students throughout the lesson. We finished the meeting talking about what the team members needed to add to their teacher edition lessons to account for each element of strong lesson design.

Through the team conversations around intentional lesson design, the team members were able to grow their collective knowledge of the essential learning standard and the pedagogical knowledge needed to support student learning.

As you begin to codesign your lesson, use the lesson-plan analysis tool in figure 8.2 to assist you with designing the lesson.

Lesson-Plan Analysis Tool

Directions: Read through the lesson plan and look for evidence to support the specific components of a well-planned lesson. Provide feedback to your colleagues to make the best lesson ever.

Lesson Element	Probing Questions	Comments, Questions, or Clarifications
Preparation for the *Why* of the Mathematics Lesson	• What are the essential mathematics learning standard and learning target or learning targets for the lesson? (What do you want students to know and understand about mathematics as a result of this lesson?) • What is the process standard of the lesson? How will students engage in the process?	
Beginning of Mathematics Class: Prior-Knowledge Routines	• In what ways does the task build on students' prior knowledge, life experiences, and culture? • What definitions, concepts, or ideas do students need to know to begin to work on the task? How will you develop students' mathematical language and notations? • What prior-knowledge routines will you use to help students access their prior knowledge and relevant life and cultural experiences? • What are the connections between important ideas in this lesson and important ideas in past and future lessons?	
Instruction: During-Class Routines (Whole-group discourse, small-group discourse, and mathematical language routines)	• What specific activities, investigations, problems, questions, or tasks will students be working on during the lesson? • How are the mathematical procedures in the lesson justified and connected with important ideas? • What are all the ways the task can be solved? • Which solution pathways will students use? • What are potential student misconceptions? • What are assessing questions or prompts you can use to get students "unstuck"? • What are advancing questions or prompts to further student understanding? • How will you engage all students to think about your questions? • How do you see or hear students engage with mathematical ideas during class? • What is the expectation of student engagement for sense making and reasoning? • How will you ensure the task is accessible to all students while still maintaining a high cognitive demand for students? • What will be the student-to-student interactions you employ? • How does the plan address whole-group and small-group discourse? • Which solution pathways do you want to share during the class discussion? In what order will you present the solutions? Why? • What opportunities exist for students to productively struggle with the mathematical ideas? • How do you respond to students' struggles, and how does your response support students to stay engaged in the mathematical tasks for the lesson?	
Closure Routines	• What are the student-led summary activities, questions, and discussions to close the lesson and provide a foreshadowing of future learning? • How will students self-assess their understanding? • How will students know that they met the learning target for the day? • What formative assessment strategies will you employ during the lesson? • What evidence will you use to determine the level of student learning of the daily learning target?	

Figure 8.2: Teacher team discussion tool—Lesson-plan analysis.

*Visit **go.SolutionTree.com/MathematicsatWork** for a free reproducible version of this figure.*

When you decide to do instructional rounds or lesson study, you are committing to making the instructional routines transparent across the team so that you can learn with and from each other.

Matt Larson reveals in his story ways that you can make your learning more visible as you help one another observe for evidence of the Mathematics in a PLC at Work lesson-design tool criteria in your own lesson-planning and implementation efforts.

Personal Story MATT LARSON

A principal asked me to observe and provide feedback concerning the classroom environment of one of her mathematics teachers. The principal was concerned because each semester, a number of students asked to transfer out of the teacher's classes. All outward signs were the teacher was highly effective. There were no discipline issues, and his colleagues valued his participation on their collaborative team, where he made significant contributions to task selection and assessment design.

I spent a day observing the teacher, and it was quickly clear what he could address to improve his classroom environment: his questioning pattern. The teacher's questioning pattern was to pose a question, leave little wait time, call on students, and then evaluate their response. His questioning pattern communicated to students that all that mattered were quick and correct answers. In addition, if students answered incorrectly, they felt demeaned by his response to them. Students simply disengaged.

He agreed to videotape his classroom, and we analyzed the tape together. We focused on the impact of his questioning pattern on how students saw themselves as learners. We worked together on his questioning approach. Over the course of the semester, he began to ask questions and listen to student responses to determine students' level of understanding, have other students respond to student responses, show respect for students' responses, and ask more open-ended questions that provided space for students' ideas. In short, he began to position students as capable doers of mathematics and, in the process, he completely transformed his classroom environment and student-led closure.

Regardless of the strategy you and your team choose to make your instructional practices more transparent, the most important step is that your team looks for ways to monitor student learning during instruction, grow teacher efficacy, and improve your and your team's instructional practices by learning *together*.

Reflection on Practice

Effectively implementing the lesson-design elements supports equitable teaching practices and outcomes. In fact, the concept of team action, the foundation of this book and the other books in the *Every Student Can Learn Mathematics* series, makes equity in learning possible by focusing the work of your collaborative team on the four critical questions of a PLC (DuFour et al., 2016). Thus, teams establish goals to focus learning and ensure that all students have access to the learning—regardless of the teacher students receive.

The goals you and your team select for your students have a profound impact on their development of a productive mathematics disposition (Smith et al., 2017). "The teacher's goals for a lesson frame the tasks that teachers choose and therefore students' opportunities to learn" (Smith et al., 2017, p. 26).

As you and your team design lessons, you may find that you have different ideas or strategies for how to plan or teach a specific standard. This is OK. When it comes to the lesson-design process, it is important to ensure everyone is clear and consistent on the use of the essential learning standards and the mathematical language as well as the level of rigor each standard requires in your selection of lower- and higher-level tasks.

However, there may be variation in the prior-knowledge, discourse, and closure routines you choose due to the formative nature of these actions; you may need to choose different activities based on the evidence of learning (or lack of evidence) your students demonstrate throughout instruction.

As a team, use figure 8.3 to reflect on the elements of the Mathematics in a PLC at Work lesson-design tool as you bring your own personal closure to the six lesson-design elements.

Directions: Closely examine the elements in the Mathematics in a PLC at Work lesson-design tool (appendix, page 111). With your colleagues, discuss how you could use parts or all of the tool to help your current efforts to design mathematics lessons for your grade level or course.		
Lesson-Design Element	**Strengths** Identify what you currently do that is a strength for each element.	**Challenges** Identify any part of the element that you currently do not address or do not address well in your lesson-designing process, and list how you might improve the element.
1. Essential learning standards: the *why* of the lesson		
2. Prior-knowledge routines		
3. Mathematical language routines		
4. Lower and higher-level-cognitive-demand mathematical task balance		
5. Mathematical discourse routines		
6. Lesson-closure routines		

Figure 8.3: Teacher team discussion tool—Protocol for team analysis of the Mathematics in a PLC at Work lesson-design tool.

*Visit **go.SolutionTree.com/MathematicsatWork** for a free reproducible version of this figure.*

Now that you and your team have reflected on the important features of a lesson and how to implement each lesson-design strategy to support student perseverance and formative assessment, visit **go.SolutionTree.com/MathematicsatWork** or scan the QR code in this book to consider an appropriate sample lesson. You will find a sample grade 1 lesson plan, a sample grade 5 lesson plan, a sample grade 8 lesson plan, and a sample high school advanced algebra lesson plan.

As you examine the sample lessons online, remember that these sample templates are to help support your lesson-design effort and focus for the mathematics content and processes of any daily lesson. Approach the tool not as something that must be completely filled in but as a guide for the types of student-engaged activities you and your team need to plan for as you create the story of each lesson. If you teach a grade above or below the samples, remember to ask yourself, "How does the lesson inform me of student work as part of a progression for student learning?"

In this book, *Mathematics Instruction and Tasks in a PLC at Work*, as authors, we have tried to help you find the right balance around the lesson-design routines and elements that have a primary impact on student perseverance in class, and most likely result in retention of learning the expected mathematics standards for your grade level or course. We also want to help you and your team learn how to use those lesson-design routines and elements more strategically each day, and how to become more confident and transparent in using those elements for your daily mathematics lessons. We hope we succeeded as you move toward greater transparency and success in your practice with colleagues.

TEAM RECOMMENDATION

Use and Reflect on the Mathematics in a PLC at Work Lesson-Design Tool

- Collaborate as a team to collectively write common essential learning standards for the unit and daily learning targets for each lesson.
- Collaborate and share ideas about the prior-knowledge routines, mathematical language routines, and lesson-closure routines.
- Collaborate and share ideas regarding the balance of mathematical tasks and whole-group and small-group discourse routines.
- After implementing the lesson, take time and reflect upon the successes of the lesson. Leave notes for the following lesson, unit, or year, and reflect on next steps in your instructional unit plan. Specifically, what might need to change in tomorrow's lesson based on student evidence of learning from today?
- Intentionally create opportunities for you and your team to make instructional routines more transparent through codesigning lessons, using instructional rounds, or engaging in lesson studies.

Visit **go.SolutionTree.com/MathematicsatWork** for free reproducible versions of tools and protocols that appear in this book, as well as additional online only materials.

Appendix

Mathematics in a PLC at Work Lesson-Design Tool

Preparing for the Lesson
Unit: Fill in the title of the unit. **Date:** Fill in the date of the lesson. **Lesson:** Provide a short descriptor about the nature of this lesson.
Essential learning standard: State the essential content and process standard *for the unit* you address during *this* lesson. • **Content**—Write as an *I can* statement. • **Process**—Write as an *I can* statement.
Learning target: State the specific learning outcome(s) for this lesson. *Use, "Students will be able to . . ."*
Mathematical language routine: State the mathematical language (both vocabulary and notations) expectations for the lesson. Describe how you will explicitly address any new vocabulary or notations.
Beginning-of-Class Routines
Prior-knowledge routine: Describe the prior-knowledge routine you will use. How does the warm-up activity connect to students' prior knowledge, connect to an analysis of homework progress, or connect to future learning?

<table>
<tr><th colspan="2">During-Class Routines</th></tr>
<tr><td colspan="2">Task 1: Cognitive Demand (Circle one): High or Low
What are the learning activities to engage students in learning the target? Be sure to list materials as necessary.</td></tr>
<tr><td>What will the teacher be doing?
• How will you present and then monitor student response to the task?
• How will you expect students to demonstrate proficiency of the learning target during in-class checks for understanding?
• How will you scaffold instruction for students who are stuck during the lesson or the lesson tasks (assessing questions)?
• How will you further learning for students who are ready to advance beyond the standard during class (advancing questions)?</td><td>What will the students be doing?
• How will you actively engage students in each part of the lesson?
• What type of student discourse does this task require—whole group or small group?
• What mathematical thinking (reasoning, problem solving, or justification) are students developing during this task?</td></tr>
<tr><td></td><td></td></tr>
<tr><td colspan="2">Task 2: Cognitive Demand (Circle one): High or Low
What are the learning activities to engage students in learning the target? Be sure to list materials as necessary.</td></tr>
<tr><td>What will the teacher be doing?
• How will you present and then monitor student response to the task?
• How will you expect students to demonstrate proficiency of the learning target during in-class checks for understanding?
• How will you scaffold instruction for students who are stuck during the lesson or the lesson tasks (assessing questions)?
• How will you further learning for students who are ready to advance beyond the standard during class (advancing questions)?</td><td>What will the students be doing?
• How will you actively engage students in each part of the lesson?
• What type of student discourse does this task require—whole group or small group?
• What mathematical thinking (reasoning, problem solving, or justification) are students developing during this task?</td></tr>
<tr><td></td><td></td></tr>
</table>

<table>
<tr><td colspan="2">Task 3: Cognitive Demand (Circle one): High or Low
What are the learning activities to engage students in learning the target? Be sure to list materials as necessary.</td></tr>
<tr><td>What will the teacher be doing?
• How will you present and then monitor student response to the task?
• How will you expect students to demonstrate proficiency of the learning target during in-class checks for understanding?
• How will you scaffold instruction for students who are stuck during the lesson or the lesson tasks (assessing questions)?
• How will you further learning for students who are ready to advance beyond the standard during class (advancing questions)?</td><td>What will the students be doing?
• How will you actively engage students in each part of the lesson?
• What type of student discourse does this task require—whole group or small group?
• What mathematical thinking (reasoning, problem solving, or justification) are students developing during this task?</td></tr>
<tr><td></td><td></td></tr>
<tr><td colspan="2">Closure Routines</td></tr>
<tr><td colspan="2">Common homework: Describe the independent practice teachers will assign when the lesson is complete.</td></tr>
<tr><td colspan="2">Lesson-closure routine: How will lesson closure include a student-led summary? By the end of the lesson, how will you measure student proficiency and students' development of a deepened (and conceptual) understanding of the learning target or targets for the lesson?</td></tr>
<tr><td colspan="2">Teacher end-of-lesson reflection: (To be completed by the teacher after the lesson is over)
Which aspects of the lesson (tasks or teacher or student actions) led to student understanding of the learning target? What were common misconceptions or challenges with understanding, if any? How should you address these in the next lessons?</td></tr>
</table>

Cognitive-Demand-Level Task Analysis Guide

This guide provides criteria you can use to evaluate the cognitive-demand level of the tasks you choose each day. As you design your lessons and unit plans, use this tool (table A.1) to ensure a balance of lower- and higher-level-cognitive-demand tasks for students as they learn the standards.

Table A.1: Cognitive-Demand Levels of Mathematical Tasks

Lower-Level Cognitive Demand	Higher-Level Cognitive Demand
Memorization Tasks • These tasks involve reproducing previously learned facts, rules, formulae, or definitions to memory. • They cannot be solved using procedures because a procedure does not exist or because the time frame in which the task is being completed is too short to use the procedure. • They are not ambiguous; such tasks involve exact reproduction of previously seen material and what is to be reproduced is clearly and directly stated. • They have no connection to the concepts or meaning that underlie the facts, rules, formulae, or definitions being learned or reproduced.	**Procedures With Connections Tasks** • These procedures focus students' attention on the use of procedures for the purpose of developing deeper levels of understanding of mathematical concepts and ideas. • They suggest pathways to follow (explicitly or implicitly) that are broad general procedures that have close connections to underlying conceptual ideas as opposed to narrow algorithms that are opaque with respect to underlying concepts. • They usually are represented in multiple ways (for example, visual diagrams, manipulatives, symbols, or problem situations). They require some degree of cognitive effort. Although general procedures may be followed, they cannot be followed mindlessly. Students need to engage with the conceptual ideas that underlie the procedures in order to successfully complete the task and develop understanding.
Procedures Without Connections Tasks • These procedures are algorithmic. Use of the procedure is either specifically called for, or its use is evident based on prior instruction, experience, or placement of the task. • They require limited cognitive demand for successful completion. There is little ambiguity about what needs to be done and how to do it. • They have no connection to the concepts or meaning that underlie the procedure being used. • They are focused on producing correct answers rather than developing mathematical understanding. • They require no explanations or have explanations that focus solely on describing the procedure used.	**Doing Mathematics Tasks** • Doing mathematics tasks requires complex and nonalgorithmic thinking (for example, the task, instructions, or examples do not explicitly suggest a predictable, well-rehearsed approach or pathway). • It requires students to explore and understand the nature of mathematical concepts, processes, or relationships. • It demands self-monitoring or self-regulation of one's own cognitive processes. • It requires students to access relevant knowledge and experiences and make appropriate use of them in working through the task. • It requires students to analyze the task and actively examine task constraints that may limit possible solution strategies and solutions. • It requires considerable cognitive effort and may involve some level of anxiety for the student due to the unpredictable nature of the required solution process.

Source: Smith & Stein, 1998. Used with permission.

Mathematics Instruction and Tasks in a PLC at Work Protocols and Tools

Team Action 3:
Develop high-quality mathematics lessons for daily instruction.

Team Action 4:
Analyze and use effective lesson-design elements to provide formative feedback and build student perseverance.

Actions Teacher teams must . . .	Questions the Protocol or Tool Will Address	Protocol or Tool
Answer questions about their current lesson design.	• How does each teacher on the team inform students of the relevance of the lesson and plan for the lesson-design elements?	Figure I.3: Teacher team discussion tool—Collaborative lesson-design elements (page 6)
Analyze and evaluate the quality of lesson design.	• Using the rubric, what is the "grade" for lesson design? • What is a strength in team lesson design? • What are areas that need to be addressed or improved?	Figure I.4: Mathematics in a PLC at Work instructional framework and lesson-design evaluation rubric (page 7) Figure 8.2: Teacher team discussion tool—Lesson-plan analysis (page 108)
Design a lesson.	• What are the key elements in a quality lesson? • How will teachers plan for the six elements of lesson design in each lesson?	Mathematics in a PLC at Work lesson-design tool (page 112)
Chapter 1: Essential Learning Standards—The *Why* of the Lesson		
Determine essential learning standards to be used for lessons and task selection in a unit.	• How do we unwrap standards to determine what students should know and be able to do in daily learning targets? • What will be the student-friendly language used for essential learning standards (ideally, five or fewer per unit)?	Sample Essential Learning Standards • Figure 1.1: Sample essential learning standards for a preK composing and decomposing numbers up to 5 unit (page 14) • Figure 1.2: Sample essential learning standards for a grade 2 applying and extending base-ten understanding unit (page 15) • Figure 1.3: Sample essential learning standards for a grade 7 statistics unit (page 16) • Figure 1.4: Sample essential learning standards for an algebra 1 linear equations unit (page 17)

Chapter 2: Prior-Knowledge Routines		
Create prior-knowledge routines.	• What are the essential learning standard and daily learning target for the lesson? • What prior knowledge is needed to learn the standard? • Which routine will you use to activate the prior knowledge?	Prior-Knowledge Routine Examples • Figure 2.1: Grade 2 prior-knowledge routine example (page 25) • Figure 2.2: Grade 5 prior-knowledge routine example (page 25) • Figure 2.3: Grade 7 prior-knowledge routine example (page 26) • Figure 2.4: High school algebra 1 prior-knowledge routine example (page 27) • Figure 2.5: Prior-knowledge routine-planning tool (page 29)
Answer questions about their current process for implementing prior-knowledge routines.	• How are prior routines currently structured and used? • How do students discuss the prior knowledge? • How do students know if their answers are correct? • What happens if students struggle?	Figure 2.6: Teacher team discussion tool—Prior-knowledge routine process (page 31)
Chapter 3: Mathematical Language Routines		
Analyze types of academic vocabulary for a unit.	• What are the academic vocabulary words students need to learn in the current unit? • What types of words are they? • Why are they a challenge?	Figure 3.1: Teacher team discussion tool—Categories of vocabulary challenges for students (page 35)
Analyze examples of mathematical language routines to use in lessons.	• What are mathematical language routines or graphic organizers teachers use that are successful? • What new activity might the team try?	Figure 3.2: Mathematical language routines (page 40) Figure 3.3: Mathematical language graphic organizers (page 41)
Share examples of academic vocabulary in a unit with a vocabulary strategy for student learning.	• How do teams determine the academic vocabulary students need to learn in a unit? • What is an example of a vocabulary strategy to use?	Figure 3.4: Teacher team discussion tool—Planning for vocabulary instruction (page 42)
Chapter 4: A Balance of Mathematical Tasks		
Identify low- and high-level-cognitive-demand tasks for essential learning standards.	• What is the difference between a low-level task and a high-level task? • What are examples of low- and high-level tasks to the same standard?	Figure 4.1: Examples of lower- and higher-level-cognitive-demand mathematical tasks (page 47) Table A.1: Cognitive-Demand-Level Task Analysis Guide (page 115)
Answer questions about their use of higher- and lower-level-cognitive-demand tasks in lessons.	• How does your team select tasks? • What percentages are of lower- and higher-level cognitive demand? • How are higher-level tasks for student feedback?	Figure 4.2: Teacher team discussion tool—Choosing mathematical tasks for lesson design during the unit (page 50)

Actions Teacher teams must . . .	Questions the Protocol or Tool Will Address	Protocol or Tool
Plan for the formative assessment process using a high-level task aligned to a standard.	• What student misconceptions are anticipated? • What are examples of scaffolding questions? • How will students get feedback? • How will students learn from one another? • How will students take action on their feedback?	Planning for the Formative Assessment Process • Figure 4.3: Planning for the formative assessment process—Kindergarten (page 52) • Figure 4.4: Planning for the formative assessment process—Grade 2 (page 52) • Figure 4.5: Planning for the formative assessment process—Grade 4 (page 53) • Figure 4.6: Planning for the formative assessment process—Grade 7 (page 53) • Figure 4.7: Planning for the formative assessment process—High school geometry (page 54) Online Only • Planning for the formative assessment process—Blank template • Additional grade-level examples
Chapter 5: Mathematical Discourse Routines		
Answer questions about student discourse in lessons.	• What percentages of a lesson are whole-group discourse and small-group discourse? • What is an obstacle to small-group discourse? • How can discourse be structured so each student is engaged in learning through the conversation?	Figure 5.1: Teacher team discussion tool—Student discourse during the lesson (page 62)
Reflect on current questioning practices.	• What questions does each teacher use? • What are strengths in questioning routines? • What feedback can teachers give one another through classroom observation?	Figure 5.2: Teacher team discussion tool—Self-evaluation checklist on whole-group questioning (page 66)
Analyze prompts used during whole-group discourse.	• What prompts can teachers use in whole-group discourse? • What prompts can students use?	Figure 5.3: Teacher team discussion tool—Whole-group discourse teacher prompts and student sentence starters (page 67)
Engage in a conversation about the differences between checking for understanding and formative assessment processes.	• What are ways to check for understanding? • What are ways to plan for formative assessment processes? • How can students get FAST feedback and take action?	Figure 5.4: Teacher team discussion tool—Checking for understanding as part of the in-class formative assessment process (page 71)

Analyze prompts used during small-group discourse.	• Which small-group discourse prompts do teachers use to elicit specific student reasoning?	Figure 5.5: Teacher team discussion tool—Assessing and advancing questions to ask students (page 72)
Use an observational tool for formative assessment.	• How can the team structure an observational tool to gather data about student learning? • How will the rubric be articulated?	Figure 5.6: Teacher team discussion tool—Observational formative assessment tool (page 74)
Share examples of classroom discourse rights and responsibilities, norms, and seating arrangements.	• What are examples of classroom agreements related to discourse? • What are examples of classroom norms? • How might seating charts be created to facilitate quality discourse?	Examples • Figure 5.7: Sample classroom discourse rights and responsibilities (page 75) • Figure 5.8: Sample classroom norms (page 75) • Figure 5.9: Seating chart to support the work of student teams during instruction (page 76)
Answer questions related to management of student teams during a lesson. Share examples of answers to questions related to management of student teams during a lesson.	• What are the routines and classroom expectations for student-to-student discourse? • What are the routines and classroom expectations for whole-group parts of the lesson? • What are possible ways to plan for productive student management?	Figure 5.10: Teacher team discussion tool—Management of student teams (page 78) Figure 5.11: Teacher team discussion tool—Management of student teams (with answers; page 79)
Share an example of a student processing and evaluation tool about collaborative teamwork.	• How can students reflect on the engagement and contribution of team members in learning?	Figure 5.12: Student team processing and evaluation tool (page 81)
Chapter 6: Lesson-Closure Routines		
Answer questions about lesson-closure activities.	• How do the members of your team plan for student-led closure? • How do teachers use evidence of learning collected from closure?	Figure 6.1: Teacher team discussion tool—Lesson-closure reflection (page 86) Figure 6.2: Teacher team discussion tool—Sample lesson-closure activities (page 90)
Chapter 7: High-Quality Tier 1 Mathematics Intervention		
Reflect on statements related to instructional practices.	• What strategies do teachers use in Tier 1 instruction? • How do the members of your team plan for differentiation?	Figure 7.1: Sample form for collecting data observations (page 95) Figure 7.2: Examples of differentiation strategies (page 96)
Chapter 8: Analyzing the Effectiveness of Mathematics Instruction		
Reflect on instructional practices.	• How does the team use the common assessment evidence to evaluate the core instruction?	Figure 8.1: Teacher team discussion tool—Connecting in-class Tier 1 intervention to the end-of-unit assessment (page 104)

Actions Teacher teams must . . .	Questions the Protocol or Tool Will Address	Protocol or Tool
Analyze lesson plans.	• What does a quality lesson design include?	Figure 8.2: Teacher team discussion tool—Lesson-plan analysis (page 107) Online Only Examples of Lessons Using the Lesson-Design Tool • Grade 1 lesson • Grade 5 geometry lesson • Grade 8 systems of linear equations lesson • Advanced algebra 2 lesson
Discuss questions related to the elements of lesson design.	• How do each teacher and the team plan for the six lesson-design elements?	Figure 8.3: Teacher team discussion tool—Protocol for team analysis of the Mathematics in a PLC at Work lesson-design tool (page 109)
Analyze instruction through the lens of a focus area using instructional rounds with feedback.	• What student actions support a team's instructional goal? • How will the teachers on the team look for the student action in classrooms? • What are examples of feedback to give one another on the team?	Online Only • Instructional rounds protocol • Instructional rounds protocol instructions
Analyze the quality of lesson design using a lesson-study protocol.	• How will team members plan a lesson collaboratively? • How will the team implement the lesson study? • How will team members collect evidence of student learning and discourse during the lesson study? • What are the time expectations for lesson study? • How will the team summarize learning and reflect?	Online Only • Sample norms for lesson study • Lesson-study protocol instructions • Lesson-study student evidence form • Time expectations of lesson study • Lesson-study team reflection

Use the QR code to access a downloadable version of this appendix and additional online resources and examples.

References and Resources

Aguirre, J., Mayfield-Ingram, K., & Martin, D. B. (2013). *The impact of identity in K–8 mathematics: Rethinking equity-based practices*. Reston, VA: National Council of Teachers of Mathematics.

Bakhshi, H., Downing, J. M., Osborne, M. A., & Schneider, P. (2017). *The future of skills: Employment in 2030*. London: Pearson.

Ball, D. L., Ferrini-Mundy, J., Kilpatrick, J., Milgram, R. J., Schmid, W., & Schaar, R. (2005). Reaching for common ground in K–12 mathematics education. *Notices of the AMS, 52*(9), 1055–1058. Accessed at www.ams.org/notices/200509/comm-schmid.pdf on June 23, 2017.

Ball, D. L., Thames, M. H., & Phelps, G. (2008). Content knowledge for teaching: What makes it special? *Journal of Teacher Education, 59*(5), 389–407.

Barton, M. L., & Heidema, C. (2000). *Teaching reading in mathematics: A supplement to* Teaching Reading in the Content Areas Teacher's Manual (2nd ed.). Aurora, CO: Mid-continent Research for Education and Learning.

Bay-Williams, J. M., & Stokes-Levine, A. (2017). The role of concepts and procedures in developing fluency. In D. A. Spangler & J. J. Wanko (Eds.), *Enhancing classroom practice with research behind principles to actions* (pp. 61–72). Reston, VA: National Council of Teachers of Mathematics.

Beck, I. L., McKeown, M. G., & Kucan, L. (2013). *Bringing words to life: Robust vocabulary instruction* (2nd ed.). New York: Guilford Press.

Ben-Hur, M. (2006). *Concept-rich mathematics instruction: Building a strong foundation for reasoning and problem solving*. Alexandria, VA: ASCD.

Benson, J. (2021). *Improve every lesson plan with SEL*. Alexandria, VA: ASCD.

Boston, M. D., & Smith, M. S. (2009). Transforming secondary mathematics teaching: Increasing the cognitive demands of instructional tasks used in teachers' classrooms. *Journal for Research in Mathematics Education, 40*(2), 119–156.

Buffum, A., Mattos, M., & Malone, J. (2018). *Taking action: A handbook for RTI at Work*. Bloomington, IN: Solution Tree Press.

Cawelti, G. (Ed.). (1995). *Handbook of research on improving student achievement*. Arlington, VA: Educational Research Service.

City, E. A. (2011). Learning from instructional rounds. *Educational Leadership, 69*(2), 36–41. Accessed at www.ascd.org/el/articles/learning-from-instructional-rounds on February 17, 2023.

City, E. A., Elmore, R. F., Fiarman, S. E., & Teitel, L. (2009). *Instructional rounds in education: A network approach to improving teaching and learning*. Cambridge, MA: Harvard Education Press.

Civil, M., & Turner, E. (2014). Introduction. In M. Civil & E. Turner (Eds.), *The Common Core State Standards in mathematics for English language learners: Grades K–8* (pp. 1–5). Alexandria, VA: TESOL Press.

College Board. (n.d.). *Sample math questions: Multiple-choice*. Accessed at https://collegereadiness.collegeboard.org/sample-questions/math on February 15, 2023.

Dale, E., & O'Rourke, J. (1986). *Vocabulary building: A process approach*. Columbus, OH: Zaner-Bloser.

Donovan, M. S., & Bransford, J. D. (Eds.). (2005). *How students learn: History, mathematics, and science in the classroom.* Washington, DC: National Academies Press.

DuFour, R. (2015). *In praise of American educators: And how they can become even better.* Bloomington, IN: Solution Tree Press.

DuFour, R., DuFour, R., & Eaker, R. (2008). *Revisiting Professional Learning Communities at Work: New insights for improving schools.* Bloomington, IN: Solution Tree Press.

DuFour, R., DuFour, R., Eaker, R., & Karhanek, G. (2010). *Raising the bar and closing the gap: Whatever it takes.* Bloomington, IN: Solution Tree Press.

DuFour, R., DuFour, R., Eaker, R., Many, T. W., & Mattos, M. (2016). *Learning by doing: A handbook for Professional Learning Communities at Work* (3rd ed.). Bloomington, IN: Solution Tree Press.

DuFour, R., DuFour, R., Eaker, R., Mattos, M., & Muhammad, A. (2021). *Revisiting Professional Learning Communities at Work: Proven insights for sustained, substantive school improvement* (2nd ed.). Bloomington, IN: Solution Tree Press.

DuFour, R., & Eaker, R. (1998). *Professional Learning Communities at Work: Best practices for enhancing student achievement.* Bloomington, IN: Solution Tree Press.

Echevarria, J., Vogt, M. E., & Short, D. J. (2010). *The SIOP model for teaching mathematics to English learners.* Boston: Pearson.

Erkens, C., Schimmer, T., & Dimich, N. (2018). *Instructional agility: Responding to assessment with real-time decisions.* Bloomington, IN: Solution Tree Press.

Fisher, D., Frey, N., & Rothenberg, C. (2011). *Implementing RTI with English learners.* Bloomington, IN: Solution Tree Press.

Frayer, D. A., Fredrick, W. C., & Klausmeier, H. J. (1969). *A schema for testing the level of concept mastery* (Working Paper No. 16). Madison: Wisconsin Research and Development Center for Cognitive Learning.

Gersten, R., Taylor, M. J., Keys, T. D., Rolfhus, E., & Newman-Gonchar, R. (2014). *Summary of research on the effectiveness of math professional development approaches* (REL 2014–010). Washington, DC: National Center for Education Evaluation and Regional Assistance.

Goodwin, B., & Rouleau, K. (2023). *The new classroom instruction that works: The best research-based strategies for increasing student achievement.* Alexandria, VA: ASCD.

Gregory, G., Kaufeldt, M., & Mattos, M. (2016). *Best practices at Tier 1: Daily differentiation for effective instruction, elementary.* Bloomington, IN: Solution Tree Press.

Hansen, A. (2015). *How to develop PLCs for singletons and small schools.* Bloomington, IN: Solution Tree Press.

Hansen-Thomas, H., Langman, J., & Farias, T. (2018). The role of language objectives: Strengthening math and science teachers' language awareness with emergent bilinguals in secondary classrooms. *Latin American Journal of Content and Language Integrated Learning, 11*(2), 193–214.

Hattie, J. A. C. (2009). *Visible learning: A synthesis of over 800 meta-analyses relating to achievement.* New York: Routledge.

Hattie, J. A. C. (2012). *Visible learning for teachers: Maximizing impact on learning.* New York: Routledge.

Hattie, J. A. C., Fisher, D., & Frey, N. (2017). *Visible learning for mathematics, grades K–12: What works best to optimize student learning.* Thousand Oaks, CA: Corwin Mathematics.

Hattie, J. A. C., & Yates, G. C. R. (2014). *Visible learning and the science of how we learn.* New York: Routledge.

Helwig, R., Rozek-Tedesco, M. A., Tindal, G., Heath, B., & Almond, P. J. (1999). Reading as an access to mathematics problem solving on multiple-choice tests for sixth-grade students. *The Journal of Educational Research, 93*(2), 113–125.

Hiebert, J. S., & Grouws, D. A. (2007). The effects of classroom mathematics teaching on students' learning. In F. K. Lester Jr. (Ed.), *Second handbook of research on mathematics teaching and learning: A project of the National Council of Teachers of Mathematics* (pp. 371–404). Charlotte, NC: Information Age.

Himmele, P., & Himmele, W. (2021). *Why are we still doing that? Positive alternatives to problematic teaching practices.* Alexandria, VA: ASCD.

Janvier, C. (Ed.). (1987). *Problems of representation in the teaching and learning of mathematics.* Hillsdale, NJ: Erlbaum.

Johnson, D. W., & Johnson, R. T. (1999). Making cooperative learning work. *Theory Into Practice, 38*(2), 67–73.

Johnson, D. W., Johnson, R. T., & Holubec, E. J. (2008). *Cooperation in the classroom* (8th ed.). Edina, MN: Interaction Book.

Jones, P. S., & Coxford, A. F., Jr. (Eds.). (1970). *A history of mathematics education in the United States and Canada* (32nd yearbook). Reston, VA: National Council of Teachers of Mathematics.

Kagan, S. (1994). *Cooperative learning.* San Clemente, CA: Kagan.

Kagan, S., & Kagan, M. (2009). *Kagan cooperative learning.* San Clemente, CA: Kagan.

Kanold, T. D., Barnes, B., Larson, M. R., Kanold-McIntyre, J., Schuhl, S., & Toncheff, M. (2018). *Mathematics homework and grading in a PLC at Work.* Bloomington, IN: Solution Tree Press.

Kanold, T. D. (Ed.), Briars, D. J., Asturias, H., Foster, D., & Gale, M. A. (2013). *Common Core mathematics in a PLC at Work, grades 6–8.* Bloomington, IN: Solution Tree Press.

Kanold, T. D., Kanold-McIntyre, J., Larson, M. R., Barnes, B., Schuhl, S., & Toncheff, M. (2018). *Mathematics instruction and tasks in a PLC at Work.* Bloomington, IN: Solution Tree Press.

Kanold, T. D. (Ed.), & Larson, M. R. (2012). *Common Core mathematics in a PLC at Work, leader's guide.* Bloomington, IN: Solution Tree Press.

Kanold, T. D., Schuhl, S., Larson, M. R., Barnes, B., Kanold-McIntyre, J., & Toncheff, M. (2018). *Mathematics assessment and intervention in a PLC at Work.* Bloomington, IN: Solution Tree Press.

Kanold, T. D., Toncheff, M., Larson, M. R., Barnes, B., Kanold-McIntyre, J., & Schuhl, S. (2018). *Mathematics coaching and collaboration in a PLC at Work.* Bloomington, IN: Solution Tree Press.

Kazemi, E., & Hintz, A. (2014). *Intentional talk: How to structure and lead productive mathematical discussions.* Portland, ME: Stenhouse.

Kenney, J. M., Hancewicz, E., Heuer, L., Metsisto, D., & Tuttle, C. L. (2005). *Literacy strategies for improving mathematics instruction.* Alexandria, VA: ASCD.

Kilpatrick, J., Swafford, J., & Findell, B. (Eds.). (2001). *Adding it up: Helping children learn mathematics.* Washington, DC: National Academies Press.

Kirschner, P. A., & Hendrick, C. (2020). *How learning happens: Seminal works in educational psychology and what they mean in practice.* New York: Routledge.

Kirschner, P. A., Hendrick, C., & Heal, J. (2022). *How teaching happens: Seminal works in teaching and teacher effectiveness and what they mean in practice.* New York: Routledge.

Kobett, B. M., & Karp, K. S. (2020). *Strengths-based teaching and learning in mathematics: Five teaching turnarounds for grades K–6.* Thousand Oaks, CA: Corwin Press.

Larson, M. R., & Kanold, T. D. (2016). *Balancing the equation: A guide to school mathematics for educators and parents.* Bloomington, IN: Solution Tree Press.

Leane, B., & Yost, J. (2022). *Singletons in a PLC at Work: Navigating on-ramps to meaningful collaboration.* Bloomington, IN: Solution Tree Press.

Liljedahl, P. (2021). *Building thinking classrooms in mathematics, grades K–12: 14 teaching practices for enhancing learning.* Thousand Oaks, CA: Corwin Press.

Marzano, R. J. (2009). *Designing and teaching learning goals and objectives.* Bloomington, IN: Marzano Resources.

Mattos, M. (2016, November 2). *Connecting PLCs and RTI* [Blog post]. Accessed at www.allthingsplc.info/blog/view/335/connecting-plcs-and-rti on May 2, 2023.

McEwan-Adkins, E. K. (2010). *40 reading intervention strategies for K–6 students: Research-based support for RTI.* Bloomington, IN: Solution Tree Press.

Miura, I. T., & Yamagishi, J. M. (2002). The development of rational number sense. In B. Litwiller & G. Bright (Eds.), *Making sense of fractions, ratios, and proportions: 2002 yearbook* (pp. 206–212). Reston, VA: National Council of Teachers of Mathematics.

Morris, A. K., Hiebert, J., & Spitzer, S. M. (2009). Mathematical knowledge for teaching in planning and evaluating instruction: What can preservice teachers learn? *Journal for Research in Mathematics Education, 40*(5), 491–529.

National Academies of Sciences, Engineering, and Medicine. (2017). *Information technology and the U.S. workforce: Where are we and where do we go from here?* Washington, DC: National Academies Press. https:/doi.org/10.17226/24649

National Board for Professional Teaching Standards. (2010). *Mathematics standards for teachers of students ages 11–18+* (3rd ed.). Arlington, VA: Author.

National Council of Supervisors of Mathematics. (2020). *NCSM essential actions: Framework for leadership in mathematics education*. Portsmouth, NH: Author.

National Council of Supervisors of Mathematics. (2022). *Culturally relevant leadership in mathematics*. Aurora, CO: Author.

National Council of Teachers of Mathematics. (1980). *An agenda for action* [Pamphlet]. Reston, VA: Author.

National Council of Teachers of Mathematics. (1991). *Professional standards for teaching mathematics*. Reston, VA: Author.

National Council of Teachers of Mathematics. (2009). *Focus in high school mathematics: Reasoning and sense making*. Reston, VA: Author.

National Council of Teachers of Mathematics. (2014). *Principles to actions: Ensuring mathematical success for all*. Reston, VA: Author.

National Council of Teachers of Mathematics. (2018). *Catalyzing change in high school mathematics: Initiating critical conversations*. Reston, VA: Author.

National Council of Teachers of Mathematics. (2020a). *Catalyzing change in early childhood and elementary mathematics: Initiating critical conversations*. Reston, VA: Author.

National Council of Teachers of Mathematics. (2020b). *Catalyzing change in middle school mathematics: Initiating critical conversations*. Reston, VA: Author.

National Council of Teachers of Mathematics & National Council of Supervisors of Mathematics. (2020). *Moving forward: Mathematics learning in the era of COVID-19*. Accessed at www.mathedleadership.org/docs/resources/NCTM_NCSM_Moving_Forward.pdf on February 10, 2023.

National Governors Association Center for Best Practices & Council of Chief State School Officers. (2010). *Common Core State Standards for mathematics*. Washington, DC: Authors. Accessed at https://ccsso.org/sites/default/files/2017-10/MathStandards50805232017.pdf on May 1, 2023.

NRICH Team. (2019). *Creating a low threshold high ceiling classroom*. Accessed at http://nrich.maths.org/7701 on February 15, 2023.

Obeng, K. (2022, January 5). *Martin Luther King Jr. said, "Education is a battleground." Reflecting on his words*. Accessed at https://ewa.org/news-explainers/martin-luther-king-jr-said-education-is-a-battleground-reflecting-on-his-words on February 16, 2023.

Panadero, E., Andrade, H., & Brookhart, S. (2018). Fusing self-regulated learning and formative assessment: A roadmap of where we are, how we got here, and where we are going. *The Australian Educational Researcher*, *45*(1), 13–31. Accessed at https://link.springer.com/article/10.1007/s13384-018-0258-y on January 21, 2023.

Pellegrino, J. W., & Hilton, M. L. (Eds.). (2012). *Education for life and work: Developing transferable knowledge and skills in the 21st century*. Washington, DC: National Academies Press.

Popham, W. J. (2011). *Transformative assessment in action: An inside look at applying the process*. Alexandria, VA: ASCD.

Rainie, L., & Anderson, J. (2017, May 3). *The future of jobs and job training*. Washington, DC: Pew Research Center. Accessed at https://pewresearch.org/internet/2017/05/03/the-future-of-jobs-and-jobs-training on February 10, 2023.

Reeves, D. (2011). *Elements of grading: A guide to effective practice*. Bloomington, IN: Solution Tree Press.

Reeves, D. (2016). *Elements of grading: A guide to effective practice* (2nd ed.). Bloomington, IN: Solution Tree Press.

Resnick, L. B. (Ed.). (2006). Do the math: Cognitive demand makes a difference. *Research Points: Essential Information for Education Policy*, *4*(2), 1–4. Accessed at www.aera.net/Portals/38/docs/Publications/Do%20the%20Math.pdf on July 10, 2017.

Riccomini, P. J., Smith, G. W., Hughes, E. M., & Fries, K. M. (2015). The language of mathematics: The importance of teaching and learning mathematical vocabulary. *Reading and Writing Quarterly*, *31*(3), 235–252. https://doi.org/10.1080/10573569.2015.1030995

Rubenstein, R. N. (2007). Focused strategies for middle-grades mathematics vocabulary development. *Mathematics Teaching in the Middle School*, *13*(4), 200–207.

Rubenstein, R. N., & Thompson, D. R. (2002). Understanding and supporting children's mathematical vocabulary development. *Teaching Children Mathematics*, *9*(2), 107–112.

Schimmer, T. (2016). *Grading from the inside out: Bringing accuracy to student assessment through a standards-based mindset.* Bloomington, IN: Solution Tree Press.

Schmoker, M. (2011). *Focus: Elevating the essentials to radically improve student learning.* Alexandria, VA: ASCD.

Schmoker, M. (2018). *Focus: Elevating the essentials to radically improve student learning* (2nd ed.). Alexandria, VA: ASCD.

Schoenbach, R., Greenleaf, C., Cziko, C., & Hurwitz, L. (1999). *Reading for understanding: A guide to improving reading in middle and high school classrooms.* San Francisco: Jossey-Bass.

Schuhl, S., Kanold, T. D., Barnes, B., Jain, D. M., Larson, M. R., & Mozingo, B. (2021). *Mathematics unit planning in a PLC at Work, high school.* Bloomington, IN: Solution Tree Press.

Schuhl, S., Kanold, T. D., Deinhart, J., Lang-Raad, N. D., Larson, M. R., & Smith, N. N. (2021). *Mathematics unit planning in a PLC at Work, grades preK–2.* Bloomington, IN: Solution Tree Press.

Schuhl, S., Kanold, T. D., Deinhart, J., Larson, M. R., & Toncheff, M. (2020). *Mathematics unit planning in a PLC at Work, grades 3–5.* Bloomington, IN: Solution Tree Press.

Schuhl, S., Kanold, T. D., Kanold-McIntyre, J., Chuang, S., Larson, M. R., & Smith, M. (2021). *Mathematics unit planning in a PLC at Work, grades 6–8.* Bloomington, IN: Solution Tree Press.

Schuhl, S., Kanold, T. D., Toncheff, M., Barnes, B., Kanold-McIntyre, J., Larson, M. R., & Rivera, G. (2024). *Mathematics assessment and intervention in a PLC at Work* (2nd ed.). Bloomington, IN: Solution Tree Press.

Silver, E. A. (2010). Examining what teachers do when they display their best practice: Teaching mathematics for understanding. *Journal of Mathematics Education at Teachers College*, *1*(1), 1–6.

Slavin, R. E. (2014). Making cooperative learning powerful. *Educational Leadership*, *72*(2), 22–26.

Smith, M. S., Steele, M. D., & Raith, M. L. (2017). *Taking action: Implementing effective mathematics teaching practices, grades 6–8.* Reston, VA: National Council of Teachers of Mathematics.

Smith, M. S., & Stein, M. K. (1998). Selecting and creating mathematical tasks: From research to practice. *Mathematics Teaching in the Middle School*, *3*(5), 344–350.

Smith, M. S., & Stein, M. K. (2011). *Five practices for orchestrating productive mathematics discussions.* Reston, VA: National Council of Teachers of Mathematics.

Smith, M. S., & Stein, M. K. (2018). *Five practices for orchestrating productive mathematics discussions* (2nd ed.). Reston, VA: National Council of Teachers of Mathematics.

Stein, M. K., Remillard, J., & Smith, M. S. (2007). How curriculum influences student learning. In F. K. Lester Jr. (Ed.), *Second handbook of research on mathematics teaching and learning: A project of the National Council of Teachers of Mathematics* (pp. 319–370). Charlotte, NC: Information Age.

Stigler, J. W., & Hiebert, J. (1999). *The teaching gap: Best ideas from the world's teachers for improving education in the classroom.* New York: Free Press.

Strickland, C. (2007). *Tools for high-quality differentiated instruction: An ASCD action tool.* Alexandria, VA: ASCD.

Tomlinson, C. A., & Strickland, C. A. (2005). *Differentiation in practice: A resource guide for differentiating curriculum, grades 9–12.* Alexandria, VA: ASCD.

Tyson, K. (2013, May 26). *No tears for tiers: Common Core tiered vocabulary made simple* [Blog post]. Accessed at https://scribd.com/doc/239941811/no-tears-for-tiers on February 15, 2023.

Van de Walle, J. A., Karp, K. S., & Bay-Williams, J. M. (2019). *Elementary and middle school mathematics: Teaching developmentally* (10th ed.). Boston: Pearson.

Vygotsky, L. S. (1978). Interaction between learning and development. In M. Gauvain & M. Cole (Eds.), *Readings on the development of children* (pp. 34–40). New York: Scientific American Books.

Walsh, J. A. (2022). *Questioning for formative feedback: Meaningful dialogue to improve learning.* Alexandria, VA: ASCD.

Wiliam, D. (2011). *Embedded formative assessment.* Bloomington, IN: Solution Tree Press.

Wiliam, D. (2016). The secret of effective feedback. *Educational Leadership*, *73*(7), 10–15.

Wiliam, D. (2018). *Embedded formative assessment* (2nd ed.). Bloomington, IN: Solution Tree Press.

Wilson, J. (2022). Initial steps in developing classroom observation rubrics designed around instructional practices that support equity and access in classrooms with potential for "success." *Teachers College Record, 124*(11), 179–217. https://doi.org/10.1177/01614681221140963

World Economic Forum. (2016, January). *The future of jobs: Employment, skills and workforce strategy for the Fourth Industrial Revolution*. Cologny, Switzerland: Author. Accessed at https://www3.weforum.org/docs/WEF_Future_of_Jobs.pdf on February 10, 2023.

Zahner, W., Pelaez, K., & Calleros, E. (2020). *Using mathematics language routines in linguistically diverse classrooms*. San Diego, CA: San Diego State University. Accessed at https://meld.sdsu.edu/wp-content/uploads/2020/05/SD-Math-Leaders-2020.pdf on February 10, 2023.

Zike, D. (2003). *Dinah Zike's teaching mathematics with foldables*. Columbus, OH: Glencoe/McGraw-Hill.

Zwiers, J., Dieckmann, J., Rutherford-Quach, S., Daro, V., Skarin, R., Weiss, S., et al. (2017). *Principles for the design of mathematics curricula: Promoting language and content development*. Stanford, CA: Stanford University. Accessed at https://ul.stanford.edu/sites/default/files/resource/2021-11/Principles%20for%20the%20Design%20of%20Mathematics%20Curricula_1.pdf on February 10, 2023.

Index

R

S

T

Mathematics Assessment and Intervention in a PLC at Work®, Second Edition
Sarah Schuhl, Timothy D. Kanold, Mona Toncheff, Bill Barnes, Jessica Kanold-McIntyre, Matthew R. Larson, and Georgina Rivera
Build collective teacher efficacy by using the Mathematics in a PLC at Work™ common assessment process. New and enhanced second edition tools support your collaborative teams in developing Tier 2 interventions as part of teacher and student reflection and action.
BKG146

Mathematics Unit Planning in a PLC at Work®, Grades PreK–2
Sarah Schuhl, Timothy D. Kanold, Jennifer Deinhart, Nathan D. Lang-Raad, Matthew R. Larson, and Nanci N. Smith
Discover how to fully answer PLC critical question one in your mathematics classroom: what do we want all students to know and be able to do? With this resource, your teacher team will acquire detailed model mathematics units, learn how to perform seven collaborative tasks, and more.
BKF964

Mathematics Unit Planning in a PLC at Work®, Grades 3–5
Sarah Schuhl, Timothy D. Kanold, Jennifer Deinhart, Matthew R. Larson, and Mona Toncheff
Part of the *Every Student Can Learn Mathematics* series, this practical resource provides a framework for collectively planning a unit of study in grades 3–5. Grade-level teams learn to work together to unwrap standards, create team unit calendars, design robust fraction units, and more.
BKF965

Mathematics Unit Planning in a PLC at Work®, Grades 6–8
Sarah Schuhl, Timothy D. Kanold, Jessica Kanold-McIntyre, Suyi Chuang, Matthew R. Larson, and Mignon Smith
Improve mathematics achievement across grades 6–8 through a collaborative unit planning process. With the authors' expert guidance, your team will learn how to collectively generate essential learning standards, create a unit calendar, identify prior knowledge, and complete many other vital collaborative tasks.
BKF966

Mathematics Unit Planning in a PLC at Work®, High School
Sarah Schuhl, Timothy D. Kanold, Bill Barnes, Darshan M. Jain, Matthew R. Larson, and Brittany Mozingo
Champion student mastery of essential mathematics content in grades 9–12. Part of the Every Student Can Learn Mathematics series, this guidebook provides high school teacher teams with a framework for collectively planning units of study.
BKF967

Visit SolutionTree.com or call 800.733.6786 to order.

Solution Tree